SYNTHETIC BIOLOGY

Safety, Security, and Promise

GIGI KWIK GRONVALL

Health Security Press
Baltimore, Maryland

An earlier version of Chapter 5. On US Leadership and Competitiveness appeared in *Health Security*, December 2015;13(6):378-389; it is reprinted here with permission of the publisher, Mary Ann Liebert Publishers, Inc.

CONTENTS

DEDICATION

For the ramp builders

FOREWORD

DAVID R. FRANZ

The homemade bumper sticker on my office wall said, *Someday, we'll look back and say, "It was fun!"* It was the mid-1990s—before a number of concerns had arisen regarding dangerous biological science, before "dual-use research of concern," before reconstruction of the 1918 flu virus in the laboratory, before H5N1 bird influenza was adapted to cause respiratory transmission in mammals, before "gain-of-function" research, before CRISPR/Cas9, and before synthetic biology had become a global revolution. It was also before September 11, 2001, and the anthrax letters that followed.

At that time, I was fighting for budget to pay salaries and keep the lights on in the laboratories of the United States Army Medical Research Institute of Infectious Diseases (USAMRIID). Researching medical countermeasures—drugs, vaccines, and diagnostics—and education to protect the warfighter against biological weapons was the mission we all understood and were devoted to. I had just lost a third of our staff positions as part of the "peace dividend" post–Cold War. Medical biological defense—and even infectious disease research on diseases other than HIV—was not high on the nation's list of priorities. My

3 top priorities in those days were safety, safety, and safety. A needlestick by a technician or a bone fragment that pierced the glove of a pathologist working at level 4, the highest level of biocontainment, could mean almost certain death. I knew a safety failure could shut us down. Safety was also an important theme around which we built a "healthy workplace culture."

Our research tools, then cutting-edge, were blunt instruments by today's standard. On the single occasion that I recall being surprised by our science and fearing we might have inadvertently produced a viral variant that could do harm if it escaped from the laboratory, I gathered a group of our best scientists and our safety officer—a fully trained and experienced microbiologist himself. Together we made the decision to slip those experimental materials quietly into an autoclave and destroy them. End of story. There was no National Science Advisory Board for Biosecurity, the federal advisory group that has weighed in on the possible intentional or accidental misuse of legitimate science. The CDC/NIH safety guidebook for biological laboratories, *Biosafety in Microbiology and Biomedical Laboratories* (*BMBL*), with its simple and time-tested guidelines regarding facilities, equipment, and procedures, was our bible. Integrity, responsibility, and common sense marked our field of play. And no one thought of publishing the unusual findings for personal or professional gain. Those days now feel like ancient history. Looking back, it *was* fun!

Synthetic Biology: Security, Safety, and Promise begins with a fascinating history of the technical foundation on which today's synthetic biological research stands. The rate of change in the tools and knowledge on which our enterprise is built has been phenomenal over the past 20 years, but a change in human behavior has also occurred. Scientific knowledge is incremental and cumulative, and in the modern age, it advances rapidly. On the other hand, human behavior and values can ebb and flow with time and change along with the health of the society from which the science must spring. I believe that while the technological tools have improved greatly, we're still struggling to find the right balance—the balance between allowing the vast majority of scientists the freedom to exploit those new tools for *good,* and keeping a small minority of careless, thoughtless, or ill-intentioned scientists from hurting themselves and the rest of us. It sounds like the evening news, but this isn't about drugs, gangs, cybercrime, guns, or intolerance; these are educated people working in legitimate and even highly respected organizations, using the latest tools of synthetic biology. However, in the relatively near future, this knowledge and some of these powerful tools will be available to the general public.

In just a decade and a half, we've gone from being focused on *laboratory worker safety*—as I was—to *laboratory security* and now even to *safety of the global community,* as we've become immersed in the silent roar of the revolution in biology going on all around us. Dr. Gronvall chronicles the government's quick reaction to the relatively few and mostly innocuous and nonlethal surprises that have come from the labs. Gigi trained and worked in a high-containment laboratory during this period of rapid change in technology, culture, and early government intervention in management of infectious disease research from inside the beltway. She has subsequently become a thought leader regarding the role of scientists in biosecurity, the position of US science in the world, and the approaches the US should take to ensure national security while promoting new advances in biotechnology. This rich background and experience make her eminently qualified to tell the synbio story.

Technologies change; fundamentals of leadership should not. Running an infectious disease research program was, in many ways, easier in the 80s and 90s than it is today. When I was given the opportunity, my boss put his arm around my shoulders and said, "You're the expert. Run the Institute as you see fit and come to me if you need help." His trust gave me significant authority, of course, but with it came a huge dose of personal responsibility. My job was now to translate the opportunity into corporate responsibility while maintaining and building an organization in which we would make technical progress. That kind of trust by senior leaders in government is almost unheard of today. When an accident or incident occurred in an infectious disease laboratory, we gathered our best experts, considered the facts, and, if warranted, changed our safety rules. Often, new practices would be adopted by the close-knit infectious disease community and ultimately incorporated into the next edition of the *BMBL*. Today, following incidents by an individual or small group, quick reactions from inside the beltway have an impact on the entire population of scientists doing related work. As a result, all scientists involved in these fields of study are now spending more time on paperwork and less in the lab, costs of the research have increased, and productivity and morale have been affected.

In the past, sound leadership, healthy cultures, and communities of trust held it all together. Today, political leadership with little or no laboratory or technical experience has reacted with layer upon layer of regulations in an attempt to control the science remotely. I don't believe it can be done. The enterprise has suffered. Dr. Gronvall describes the creeping incremental regulation of synthetic biology and related disciplines that have some experts concerned about our own global competitiveness. She also describes some great examples of very responsible self-regulation and stewardship that have been driven from within the synbio community.

Since I left the lab almost 20 years ago, I have worked across the globe using science as a common language to chip away at walls of mistrust and sometimes ignorance between nations: making friends with science. I have met exceptional individuals and visited centers of excellence in many parts of the world, but by several measures the capability and productivity of the US biotech enterprise broadly has remained impressive and dominant. The strong basic science that has come from our academic and government labs has contributed enormously to advanced development and production in the health, food, and agricultural sectors of other nations. In the past 10 years, I've also seen advances in the quality of the science, particularly from the developing world. Many of their best scientists trained in the US, Canada, the UK, or Europe before returning to their home nations.

Today, I still see unwavering respect for our best individual scientists, but also the sense that we are no longer invincible as a national enterprise. Finally, here at home, I talk to senior infectious disease scientists who say the regulatory and paperwork burden is grinding them down and slowing their progress significantly. The motivation of regulators is probably mostly genuine, but the barnacles of bureaucracy that follow have made research harder, slower, and more expensive. The layers of new regulations are said to have had some positive impact on laboratory safety—although there had been very few laboratory-acquired illnesses and I know of only 1 infectious disease laboratory death in the US over the preceding 30 years. The impact of new regulations on security is nearly impossible to measure, but increased micromanagement of our labs does not improve cultures of responsibility or encourage communities of trust. Enlightened leadership should be rewarded, not hobbled.

Finding the balance between safety/security and productivity hasn't been easy. I'm convinced we haven't found the sweet spot so critical to our remaining competitive in a globalized, biotech-driven economy. Gigi has nicely captured the complex subtleties of this fast-moving story, whose outcome has strategic implications for our health, economy, and even our national security.

Synthetic Biology: Safety, Security, and Promise illuminates the powerful tools and knowledge that have come from synthetic biology and their global spread. It chronicles human behaviors in the field and government reactions in the US and abroad. It illustrates the dramatic trade-offs between these variables in a world dominated by the 24-hour news cycle. All of us—scientists, citizens, and government officials—would do well to learn, listen, and think about the implications for our future.

David R. Franz, DVM, PhD was Commander, United States Army Medical Research Institute of Infectious Diseases.

CHAPTER 1.

ON THE NEW WORLD OF SYNTHETIC BIOLOGY

Every day is a biology news day. There are promising medical treatments, amazing facts, disputed facts, and ethical dilemmas served up in each day's headlines. And there is the business side of biology, as well, with billions of dollars at stake in patent wars, pharmaceutical deals, clinical trials, and biotech mergers. But in spite of the daily bio-news feast, it is easy to question whether we are truly living in an "Age of Biology," as the 21st century is so often described.[1-3] Despite the availability of more treatment options than there were a decade ago for numerous chronic and infectious diseases, it may be hard to see concrete evidence of biology becoming more integrated into people's lives, like smart phones have, or the internet. But in fact, biology has been stealthily taking root, yielding important shifts in manufacturing and medicine that, if anything, exceed the hype about a new biological age. Biology is now a significant force in the US economy, and it poses both immense opportunities and risks to national security. This state of affairs would not be possible without simultaneous advances in other technical fields in addition to biology—particularly engineering, computer science, and chemistry—which have converged to integrate biology into our daily lives. The convergence of these technical fields, which are poised to shape our biological century, has become known as "synthetic biology."

Precise definitions for the synthetic biology field have been difficult to construct.[4] One group of synthetic biology researchers describe the area as (1) "the design and fabrication of biological components and systems that do not already exist in the natural world" and (2) "the re-design and fabrication of existing biological systems."[5] Additionally, the synthetic biology field includes the manufacturing of standard biological parts, such as measurement devices, inverters, and logic devices, that can be engineered into genetic machines; determining the conditions for the emergence of life using protocells; and the production of cell factories for biofuels or other industrially important compounds.[6]

However, most people define synthetic biology in terms of what the field aims to do: to make biology easier to engineer.[4,7,8] While the synthetic biology field is not the first instance of biology being declared "engineerable," the increased power of the tools available to synthetic biologists make this field different: It's been described as "genetic engineering on steroids."[9,10] The ability to fruitfully accomplish bioengineering is much greater now than it has been at any other point—certainly more than when French biologist Stephane Leduc wrote about the concept in the early 1900s, or in 1974, when the renowned cancer biologist Waclaw Szybalski described then-current work on molecular biology as the "descriptive phase," adding that the real challenge will begin "when we enter the synthetic biology phase of research in our field. We will then devise new control elements and add these new modules to the existing genomes or build up wholly new genomes."[11,12(p23)] Synthetic biology is a new field and a new phenomenon thanks to advances in computing power; the ability to rapidly, reliably, and inexpensively synthesize long tracts of DNA; new tools to reliably edit genomes; an increased understanding of biological systems; and the enthusiasm of young scientists who want to enter the field.

Synthetic biology has spurred new applications in a variety of industries, including medicine, agriculture, and manufacturing. The economic impact is thought to be vast: Fidelity Investments describes synthetic biology as "the defining technology of the next century" for global investments, and in 2012, the World Economic Forum ranked synthetic biology as the second key technology for the 21st century, after informatics; in 2016, it is still listed as among the top 10 emerging technologies.[13-15]

Yet, while synthetic biology promises vast opportunities in economically and strategically important sectors, there are also risks. This book is dedicated to a discussion of the risks that are inherent in synthetic biology and the policy options that can be put into place to minimize risks and maximize the benefits of these advances. While the risks of synthetic biology are global, and the practice of science and synthetic biology is undertaken worldwide, this book focuses on policy options that are available for the United States to take in response to the risks and opportunities of synthetic biology, as well as the consequences to US national security if actions are not taken.

These are still early days for synthetic biology, with many surprises in store; synthetic biology has made impressive inroads in a variety of sectors of economic and national security importance, but there is room to grow in the years ahead. If synthetic biology follows the pattern of other powerful technologies, the surprises will be both good and bad. Synthetic biology will require the United States to adopt policies to grow the field responsibly, to compete with other nations for resources and economic gain, and to set the rules of the road to govern the technology. Minimizing risks will be a challenge for global governance, as the technology is by no means exclusively a US phenomenon. China in particular has made synthetic biology investments a priority; it is seen as a future source of economic growth.[16,17] Meanwhile, there are well-documented concerns that the United States is falling behind in biotechnology and in the sciences generally.[18-20]

Synthetic biology is an academic field of study, but it also encompasses a suite of powerful tools that can be used beneficently or can be misused by individuals, groups, or companies. It can be scaled up or scaled down. It can be practiced on a large, corporate level, involving hundreds of scientists working together on a single product, or it can be practiced by an individual who may not have much expertise. On the large scale, synthetic biology is accelerating a trend toward the industrialization of biology, where the use of biological tools is replacing the use of petrochemicals in manufacturing a variety of products, as well as replacing resource-intensive agricultural processes. But synthetic biology is also accelerating a trend toward the personalization of biology, wherein individuals can use synthetic biology for their own purposes—which could be benign, wonderful, profitable, or harmful.

To set the stage for a consideration of the risks in synthetic biology, this chapter describes the synthetic biology landscape: what synthetic biology has been used for, who practices it, the trends of industrialization and personalization that synthetic biology is accelerating, and why the United States needs to adapt its policies to address the new risks posed by the technology. Chapter 2 focuses on the issue of biosecurity, how to prevent the misuse of synthetic biology for biological weapons, and of the use synthetic biology for biodefense. Chapter 3 delves into biosafety in synthetic biology and how to prepare the field of biosafety for a time when science is increasingly performed outside of a traditional laboratory environment. Chapter 4 focuses on the ethics and public engagement of synthetic biology and why these subjects need to always be part of the scientific development of the field and can never be considered to be "done." Finally, Chapter 5 discusses why the United States must remain competitive in synthetic biology and what is at stake for the development and governance of this technology. In addition to the economic factors, there is the question of governance: If the US is not on the leading technical edge of synthetic biology advances, it will be disadvantaged in setting rules and common practice for the governance of the technologies. This will be critical for developing the safety mechanisms needed for applications such as, for example, biological control of mosquitos and other pests, or for deliberate modification of the human genome to avoid disease.

On Industrialization

Synthetic biology is accelerating an important trend in biological research: the industrialization of biology. Of course, biology has been used in industrial processes for a very long time—in agriculture, for example, as well as some medicines, vaccines, and consumer products like beer and wine. What is different now is that biological processes are increasingly used for manufacturing, replacing the use of petrochemicals and resource-intensive harvesting from nature. This expands the biological footprint for a range of industries, including fuel, agriculture, medicine, mining, construction materials, perfumes, fibers, and adhesives.[21]

The first major synthetic biology product was a treatment for malaria: the synthetic precursor of artemisinin. Artemisinin, derived from the *Artemisia annua* plant, is recommended by the World Health Organization for treating malaria and has been used for medicinal purposes since at least 200 BC. However, artemisinin has often been in short supply, as the amount grown by farms in China, Vietnam, and Kenya have been inadequate to the demand, leading to fluctuating availability.[22] Jay Keasling's laboratory at the University of California, Berkeley, synthetically recreated the biochemical pathway that produces this life-saving drug in Baker's yeast, starting with work published in 2003, and more than a decade later, synthetic artemisinin is being produced and used in sub-Saharan Africa.[23-28]

Another example of replacement of resource-intensive harvesting comes from the synthetic production of vanillin, the most dominant flavor compound in vanilla extract. Vanilla extract comes from vanilla beans, which are highly labor intensive to grow; to produce the beans commercially, the orchids that produce them need to be pollinated by hand.[29] Vanilla is the second most expensive spice after saffron. Currently, the global demand for vanilla flavoring can't be satisfied by vanilla beans alone, so most vanillin is made artificially, either from petrochemicals or from chemically treated paper pulp.[30] However it is made, vanillin does not taste quite as good as natural vanilla extract, because it is just one component of the vanilla flavor—the vanilla bean has more than 250 flavor and aroma compounds in addition to vanillin that give it its complex flavor.[30] But there are definite advantages in using synthetic vanillin, because synthetic production is not affected by weather or crop failures, which results in a steady supply and less price volatility.[31] The Swiss synthetic biology company Evolva has made synthetic vanillin and has also developed a synthetic version of saffron, partnering with International Flavors & Fragrances (IFF-USA) to produce it.[30] Interestingly, vanillin can be labeled as "natural" because it is made from the natural process of fermentation—or "brewing"—which differentiates it from vanilla derived from petrochemicals.[31,32]

Tires are yet another example of replacement by synthetic biology: They used to be entirely made from natural rubber, harvested from *Hevea* trees found in Thailand, Indonesia, and Malaysia. Again, the supply is insufficient to meet extraordinary demand: In 2015 alone, 1.7 billion tires were produced worldwide.[33,34] To fill the gap, tire manufacturers use synthetic rubber (isoprene) made from petroleum, and this accounts for 70% of the supply used for tires.[33] Approximately 7 gallons of oil are required to make 1 tire: 5 as feedstock, and 2 to supply the energy involved in manufacturing.[33] Recently, several companies have been seeking bio-based alternatives that are not dependent on fluctuating oil prices. Goodyear partnered with Dupont (which had subsumed the biotech company Genencor), and Michelin has partnered with Amyris to bring BioIsoprene to market.[35-38] As of this writing, prototype tires have been developed but are not yet available for sale.[39,40] Dupont has brought to market other products with synthetic biology, such as Dupont Sorona® stain-resistant fibers, which are used in everything from carpets to car interiors and have been in production since 2006.[41]

Other applications for synthetic biology are being explored in a diverse range of economic activity. Products like detergents and construction materials are bringing synthetic biology into people's lives, whether or not they know the products' origins. Universal BioMining is using synthetic biology for mining, to increase the efficiency of traditional mining methods for gold and copper.[42] Synthetic silk is on its way to market. Silk is known to be biodegradable and hypoallergenic; it has tensile strengths comparable to steel and may be even tougher than Kevlar, and yet it is soft and durable enough to go in the washing machine.[43-45] A National Science Foundation–funded start-up called Bolt Threads is using a fermentation process to produce spider silk proteins in yeast and recently signed a $50M deal with Patagonia to use the silk in their products.[44,45]

Biofuels have thus far been difficult for synthetic biology, as the margins are razor thin, the volumes are large, and the prices are buffeted by policy and politics, leading companies to seek alternatives. One synthetic biology company, Solazyme, actually shifted focus from biofuels to cosmetics. They produce the

Algenist skin care line, sold by Sephora, and their website describes how what "began as research for alternative renewable energy sources evolved into one of the biggest skincare breakthroughs in anti-aging technology harnessed from Microalgae."[46] Still, there are major investments in biofuels, particularly by the US Department of Energy, such as the Joint BioEnergy Institute in Emeryville, California, which aims to bring the "sunlight to biofuels pipeline" together.[47] The market and price points for biofuels may one day be more favorable.

On Tools

In addition to its commercial success, the synthetic biology field has produced biological engineering tools that have been widely adopted throughout the biological sciences. For example, DNA synthesis, or the ability to string A's, T's, C's, and G's together to "write" DNA, was limited to perhaps 30 or 40 base pairs in the 1990s. By 2010, researchers from the J. Craig Venter Institute had synthesized the 1.08 million base pair *Mycoplasma mycoides genomere* bacterial organism and "booted it up" to make a replicating cell. Now, genetic information that encodes genes can be ordered online, at approximately 10 cents/base pair, with a 10-day turnaround time, which is a price point and timeframe that is only likely to drop. The ability to write genes makes experimentation with synthetic genes an affordable, accurate, and quick option for laboratories developing new products or conducting basic research.[48,49]

Another synthetic biology tool that has permeated the biological research world is CRISPR/Cas9, a gene editing tool that allows sections of DNA to be cut and pasted as in a Word document.[50] Applications have spread far beyond the synthetic biology world to other areas of biology and have opened up possibilities that could not have been imagined a decade ago. For example,

CRISPR may one day be used to cure HIV, as it has been used in animal models to edit the virus out of latently infected cells.[51] CRISPR has also touched off ethical discussions—for example, whether it should be used to edit the human germline, making genetic changes that can be passed down from parents to children.[52-54] It is theoretically possible to use CRISPR to permanently cure inherited diseases, such as the serious blood disorder beta thalassemia, but it would also be possible to use it to produce "designer babies" with specifically desired traits. CRISPR could also be used to make a "gene drive." Typically, only half of offspring inherit a specific gene from a parent, but with a gene drive, almost all offspring would inherit that gene. A gene drive could theoretically be used eliminate a species of mosquito that could harbor the dangerous Zika, dengue, or Chikungunya viruses—but there is a clear potential for misuse of a gene drive for a biological weapon, or for unsafe applications with far-reaching consequences that may be difficult to reverse.[55-58]

Fundamental research for synthetic biology is still needed. This is largely undertaken by university researchers who specialize in synthetic biology around the globe, as well as research consortia like Synberc, funded by the US National Science Foundation.[7] As a result of continued foundational research, there are likely to be more tools developed over time. For example, there is growing research interest in developing synthetic biology sensors, not only for sensing environmental stimuli like arsenic in the water, but sensors that can detect how well a synthetic organism is producing a desired product. Sensors like these would be used as a screening mechanism to select for commercially valuable organisms and would considerably shorten the design-build-test cycle to develop new products.[59-61]

Synthetic biology is also opening new possibilities for information storage, an area that is also being explored by Microsoft in an effort with synthetic biology company Twist Bioscience. DNA is very stable and can be "read" after thousands of years, making it potentially useful for long-term archiving of information.[62,63]

The Personalization of Biology

Any uneasiness about the extraordinary power and scope of synthetic biology technologies in the hands of individuals is not helped by the blinding speed of the democratization of these technologies. Even though CRISPR is only a few years old, it has been developed into an affordable kit that amateur scientists in the "DIY Bio," or citizen science, movement can use. One CRISPR kit was crowd-funded on Indiegogo and produced by a synthetic biologist, Josiah Zayner, who wanted to make scientific advances more accessible. Computers changed his life when he was growing up on a farm in Indiana, and so he wanted to replicate that experience for people who do not have access to specialized training and equipment.[64] He was quoted as saying, "For me, it's all about 'how do we give those geniuses who are sitting in their moms' basements access, so they can do brilliant things?'"[64] The speed with which CRISPR has been democratized is highly unusual. Typically, it takes years before cutting-edge research tools developed in the most cutting-edge laboratories are found in established laboratories around the world—and much longer than that before those tools can be adapted for use in someone's kitchen. There are other kits available to enhance individual exploration of synthetic biology; for example, the company Synbiota aims to "standardize biotechnology, making it accessible to everyone."[65] They sell "tinker kits" that include software and "wetware," so that purchasers can use simple biological parts like Legos to create simple genetic circuits, to make bacteria glow, or to create their own biological inventions.

This interest in doing synthetic biology either at home or in a community laboratory exemplifies another trend in the biological sciences that is accelerated by synthetic biology: the *personalization* of biology. Biology is now being used by individuals for purposes that are useful to them in their daily lives. Someone without years of experience can use synthetic biology for purposes that will never be the subject of a traditional National Institutes of Health (NIH) research grant, such as to learn medical or genetic information, or to detect toxins in food or the presence of poisonous melamine in baby formula.[66] Using easy-to-follow techniques, a DIY Bio practitioner could determine whether the expensive fish he or she bought was actually the advertised species or a cheaper one, or determine which owners could be held responsible for not picking up after their dogs.[67,68] An amateur project could be less useful and more whimsical, such as making a bacterial culture smell like bananas, or it can be social commentary expressed in biological art.[69,70]

Perhaps the best demonstration of the personalization of biology is the International Genetically Engineered Machine competition, or iGEM. It began as a small class at MIT in Cambridge, Massachusetts, in 2003, and in 2015 it had more than 2,000 international participants and more than 16,000 alumni.[71] In iGEM competitions, undergraduates from around the world form teams and are given a kit of standard biological parts called BioBricks. Over one summer, and with the help of 2 instructors, the teams use the parts they were given and others they create to engineer biological systems and operate them in living cells. The team projects have been ambitious and sophisticated, even though many of the students are entirely new to bioscience. Many tackle real-world problems, such as bacteria that functions as an arsenic sensor, or a bacteria-produced blood substitute that can be stored for long periods. One team project from Imperial College, London, led to the creation of a synthetic bacteria that can produce cellulose, which expands the commercial use of the material—and which may someday be used for water filters that can filter out particular contaminants or a wearable fabric with sensing capabilities. The group named their synthetic bacteria *Komagataeibacter rhaeticus iGEM (K. rhaeticus)* in honor of the competition.[72,73]

The iGEM teams' accomplishments are judged at regional levels and culminate in a world championship jamboree, where the winning team gets to bring home the trophy, a large aluminum Lego, until the next year's competition.[74] Randy Rettberg, who founded the competition at MIT along with Tom Knight, Gerald Sussman, and Drew Endy, and is president of the now-independent iGEM Foundation, says they deliberately compete for students' attention with organized sports and cultivate necessary and modern skills in team-building and problem solving.[75]

Outside of academic settings, synthetic biology has caught on in the citizen science or "DIY Bio" movement. Community laboratories have popped up in Brooklyn, Boston, Seattle, San Francisco, and Baltimore—as well as in Budapest, Manchester (UK), Munich, Paris, and Prague, with many other DIY communities without laboratory space all over the world.[76] Because of restrictions in Europe on genetic manipulations, the kinds of "biohacking" gaining popularity in the United States are not always possible (that is, without a license from the government). Some companies, like Synbiota, are trying to change that, pushing for DIY Bio to be an allowable activity under the law.[77] Some are concerned that continuing to prohibit it will only push the activity underground, where it will be harder to detect misuse and where it would not be exposed to safety programs.[78]

Many community laboratories offer beginning synthetic biology classes to the public and make available space for members to pursue projects of interest. As a result, ordinary people now have access to a wide array of scientific services. DIY Bio using synthetic biology has made it to South by Southwest (SXSW), an annual set of film, music, and media conferences held in Austin, Texas, with demonstration bio projects. Joi Ito, the director of the MIT Media Lab, was quoted at SXSW as saying, "In two days, you are able to design a gene sequence, assemble it, stick it into a bacteria, and reboot the thing. . . . A few years ago this would be Nobel Prize–winning stuff. Now you can do it in a kitchen."[79]

In addition to individual exploration, community laboratories work on amateur group projects. The Boston community laboratory, called "BosLab," has an ongoing group DIY Bio project to understand what goes into the truffle.[80] Truffles are fungi that look like potatoes, grow underground, and are the most expensive food item in the world—they can cost $3,600 for a pound.[81,82] The flavor and scent of truffles comes not just from the fungus itself but from associated microbes, mostly bacteria, that are mixed in with the fungi.[82,83] BosLab plans to isolate the truffle microbes, identify them via sequencing and microscopic observation, and cultivate each one individually to understand how the natural truffle is created.

Another community laboratory, based in San Francisco, is working to produce a vegan cheese—that is, a cheese that is made in the laboratory and does not use any animal products. When the project was put on the crowdfunding site Indiegogo, $37,369 was raised by 696 backers, overshooting their fundraising goal of $15,000 by 249%.[84] To create the proteins involved in making cheese, they study animal genomes for milk-protein genetic sequences and put those genetic sequences into yeast; the cellular machinery then takes over to produce real milk-protein. They advertise that not only will no animal proteins be involved in the making of vegan cheese, but "the purified proteins will be identical to those found in regular cheese, and will not contain any [genetically modified organisms] GMO!"[84]

On Militarization

In the United States, most of the funding for synthetic biology research has come from the National Science Foundation and the Department of Energy. In general, NIH has seen the field as an engineering discipline outside of their funding remit.[85] The Department of Defense (DoD), however, has provided a substantial amount of money for synthetic biology development. A report evaluating the contribution of this work to defense stated that the field has the potential "to have major effects on commodity and specialty materials, sensing, human performance, medical, and biological and chemical weapons threats and defense, all of which are of substantial importance to the U.S. Department of Defense (DoD)."[86(piii)] The Defense Advanced Research Projects Agency (DARPA), the agency credited with the development of the internet, has made substantial investments in synthetic biology, much of it foundational to the field. One of DARPA's programs, called "Living Foundries: Advanced Tools and Capabilities for Generalizable Platforms," was able to generate major advances in the production of synthetic organisms, including a more than 7-fold acceleration and 4-fold decrease in costs between 2012 and 2014.[86] DARPA is also funding an ACE: Advanced Capabilities in Engineering Biology project that seeks to create "new chemicals and materials, sensing capabilities, therapeutics, and numerous applications."[87]

The involvement of the DoD in funding synthetic biology has led to some questions in the community about the militarization of the technology. For example, DoD has funded the use of synthetic biology to produce a safer, more environmentally friendly method to make explosives.[85] As another example, the Office of Naval Research is funding synthetic biologist Jim Collins at Boston University to develop "microbiorobots"—bacteria able to sense materials in the environment, such as underwater mines.[85] This may be just the beginning: A DoD report critiques the department for not fully embracing the technologies even more, and it recommends instituting bioengineering programs at each defense academy, requiring students to spend a summer at a major research university and funding the students' participation in the iGEM competition.[86]

Security Implications

The democratization of synthetic biology and synthetic biology tools necessarily increases the risk and consequences of misuse, especially for the development of biological weapons. It is important to note that there is no requirement for synthetic biology or other new technical advances in order to misuse biology or biological organisms. The risk of biological weapons use has been present for a long time: The pathogens that are currently regulated by the United States due to security risks—including *Bacillus anthracis,* which causes anthrax disease, and *Franciscella tularensis,* which causes tularemia—are found in the environment and in laboratories and infect people and animals all over the world.[88] However, synthetic biology does increase risks for biological weapons use. Without synthetic biology, a bioweaponeer would need to acquire a pathogen from nature or a laboratory to be used as a weapon. With synthetic biology techniques, it is possible that a pathogen could be made "from scratch," based on the genetic sequence of the pathogen, available on the internet. Some difficult-to-access pathogens could also theoretically be made this way, including smallpox virus, which was declared eradicated from the natural world in 1980.[89]

There are safety risks in synthetic biology. For some applications, synthetic organisms are not intended to be contained in the laboratory but deliberately released into the environment:

mosquito control, agriculture, pollution remediation—including from water, mining, biofuels, and medications—and even the re-creation of extinct animals like wooly mammoths.[90,91] For these "outside the laboratory" applications, there could be unintended and accidental consequences if biosafety risks are not addressed and carefully thought through. Also, practitioners of synthetic biology may not have years of experience and training in laboratories, where biosafety concerns are immediate, as is the case in infectious disease laboratories, and may not know how to take correct precautions to protect themselves and others. This is especially a risk for amateur or DIY Bio scientists, who could theoretically find themselves in safety situations that outmatch their experience in containment. Finally, it is concerning that because of the increased access to powerful technologies for more laboratories and amateurs in the world, bio-errors may occur more frequently. If an accident occurs with a transmissible pathogen, the consequences could spread well beyond the laboratory.

The ethics of synthetic biology and the public embrace of the tools are another contentious issue; synthetic biology has been recognized to have ethical and societal implications since the inception of the field. Building on receptivity of the synthetic biologists to work with social scientists, the US government, the European Commission, the UK government, and private foundations such as the Alfred P. Sloan Foundation provided direct funding for the ethical, legal, and societal implications of synthetic biology. Together, this funding created an international community of individuals informed about synthetic biology, who are aware of the applications in progress and on the horizon and who could productively engage with policymakers, the press, and the public about new synthetic biology developments, risks, and benefits. While funding in this area has since dropped off, it would benefit the entire synthetic biology field to maintain that informed community, who can ensure that the synthetic biology field advances in the public interest and that that interest is communicated to the public.

Finally, there is the risk that the United States will fall behind in the development of synthetic biology. The prospect that synthetic biology and related technologies could become the manufacturing base of the 21st century has not been lost on many nations around the world, some of which have developed plans specifically to grow their bioeconomies to take advantage of the growing economic importance of the biosciences. Some, including the UK, India, and China, have developed roadmaps intended to guide and catalyze investments in synthetic biology.[16] Given the increasingly competitive international environment, the US lead will inevitably shrink, and the United States risks falling behind.[92,93]

If the United States were to lose its competitive edge in synthetic biology and related technologies, there would be serious consequences for national security. Some negative effects would be strictly economic, resulting in a declining environment for businesses and workers to be productive in synthetic biology–related industries in the long term.[19] This is important for national security because, as described in the US *National Security Strategy* (2015), "In addition to being a key measure of power and influence in its own right, [a strong economy] underwrites our military strength and diplomatic influence. A strong economy, combined with a prominent U.S. presence in the global financial system, creates opportunities to advance our security."[94(p15)] Losing competitiveness in synthetic biology could also limit specific security applications on the horizon that are essential for national defense, such as the development of medical countermeasures for responding to biological, chemical, or radiological weapons threats and new approaches to diagnostics. Current forecasting would suggest that a loss of economic opportunities in synthetic biology could be immense: The global synthetic biology market reached $2.1 billion in 2012 and $2.7 billion in 2013. The market is expected to grow to $11.8 billion in 2018.[95]

In summary, synthetic biology is a fast-moving, complicated field that is on track to make this century—as advertised—the Age of Biology. The tools of synthetic biology are democratized, used by corporations, community laboratories, and individuals. Consequently, there are a wide variety of applications being developed. Industries are using it to make their products, and people without traditional scientific training are using it to better understand their medical conditions or for fun. While there is tremendous positive potential in the pursuit of synthetic biology through industrialization and personalization, there are also risks that need to be continually monitored. Biosecurity, biosafety, ethics, public engagement, and US competitiveness concerns need to be specifically addressed by the US government, policymakers, and synthetic biologists themselves, to minimize the risks inherent in synthetic biology and to safeguard its enormous promise.

References

1. Stavridis J. The dawning of the age of biology. *Financial Times* January 19, 2014. http://on.ft.com/2dq1gNx. Accessed August 11, 2016.
2. Glover A. The 21st century: the age of biology. OECD Forum on Global Biotechnology; November 12, 2012; Paris. https://www.oecd.org/sti/biotech/A%20Glover.pdf. Accessed August 11, 2016.
3. Rifkin J. This is the age of biology. *Guardian* July 28, 2001. http://www.theguardian.com/politics/2001/jul/28/highereducation.biologicalscience. Accessed August 11, 2016.
4. Synthetic Biology Project. What is synthetic biology?

Woodrow Wilson International Center for Scholars. http://www.synbioproject.org/topics/synbio101/definition/. Accessed May 27, 2016.
5. syntheticbiology.org. http://syntheticbiology.org/FAQ.html. Accessed May 27, 2016.
6. Paddon CJ, Westfall PJ, Pitera DJ, et al. High-level semi-synthetic production of the potent antimalarial artemisinin. *Nature* 2013;496(7446):528-532.
7. Synberc. What is synbio? https://www.synberc.org/what-is-synbio. Accessed May 27, 2016.
8. Joyce S, Mazza A-M, Kendall S, rapporteurs; Committee on Science, Technology, and Law; Policy and Global Affairs; Board on Life Sciences; Division on Earth and Life Studies; National Academies of Engineering; National Research Council. *Positioning Synthetic Biology to Meet the Challenges of the 21st Century: Summary Report of a Six Academies Symposium Series.* Washington, DC: National Academies Press; 2013.
9. Voosen P. Synthetic biology comes down to earth. *Chron High Educ* March 4, 2013. http://chronicle.com/article/Synthetic-Biology-Comes-Down/137587/. Accessed August 11, 2016.
10. Benner SA, Yang Z, Chen F. Synthetic biology, tinkering biology, and artificial biology. What are we learning? *Comptes Rendus Chimie* 2011;14(4):372-387. http://www.sciencedirect.com/science/article/pii/S1631074810001852. Accessed August 11, 2016.
11. Leduc S. *Théorie physico-chimique de la vie et générations spontanées*. Paris: A. Poinat; 1910. http://www.biodiversitylibrary.org/item/76908. Accessed August 11, 2016.
12. Szybalski W. In vivo and in vitro initiation of transcription. In: Kohn A, Shatkai A, eds. *Advances in Experimental Medicine and Biology, Control of Gene Expression*. New York: Plenum Press; 1974.

13. Synthetic Biology—Fidelity Investments. *Washington Post* website. 2012. http://www.washingtonpost.com/business/sponsored-content-synthetic-biology—fidelity-investments/2012/05/02/gIQAvIgPwT_video.html. Accessed October 11, 2015.
14. Global Agenda Council on Emerging Technologies. The top 10 emerging technologies for 2012. World Economic Forum website. February 15, 2012. http://forumblog.org/2012/02/the-2012-top-10-emerging-technologies/. Accessed August 12, 2016.
15. Cann O. These are the top 10 emerging technologies of 2016. World Economic Forum website. June 23, 2016. https://www.weforum.org/agenda/2016/06/top-10-emerging-technologies-2016/. Accessed August 12, 2016.
16. Organisation for Economic Co-operation and Development (OECD). *Emerging Policy Issues in Synthetic Biology.* Paris: OECD Publishing; 2014. http://www.oecd-ilibrary.org/science-and-technology/emerging-policy-issues-in-synthetic-biology_9789264208421-en. Accessed August 12, 2016.
17. Sender H. Chinese innovation: BGI's code for success. *Financial Times* February 16, 2015. http://on.ft.com/2dohvI0. Accessed August 12, 2016.
18. National Institutes of Health. Global competitiveness—the importance of U.S. leadership in science and innovation for the future of our economy and our health. February 2015. https://report.nih.gov/UploadDocs/nih_impact_global.pdf. Accessed August 12, 2016.
19. Porter M, Rivkin J. What Washington must do now: an eight-point plan to restore American competitiveness. *The Economist* November 21, 2012. http://www.hbs.edu/competitiveness/Documents/theworldin2013.pdf. Accessed August 12, 2016.
20. U.S. science and technology leadership increasingly

challenged by advances in Asia [press release]. National Science Foundation. January 19, 2016. http://www.nsf.gov/nsb/news/news_summ.jsp?cntn_id=137394&org=NSB&from=news. Accessed August 12, 2016.
21. Synthetic Biology Project. Inventory of synthetic biology products—existing and possible. Woodrow Wilson International Center for Scholars. July 27, 2012. http://www.synbioproject.org/process/assets/files/6631/_draft/synbio_applications_wwics.pdf. Accessed August 12, 2016.
22. Small farmers cash in on Artemisinin production. Irin website. January 21, 2009. http://www.irinnews.org/report/82486/kenya-small-farmers-cash-artemisinin-production. Accessed May 26, 2016.
23. Martin VJ, Pitera DJ, Withers ST, Newman JD, Keasling JD. Engineering a mevalonate pathway in Escherichia coli for production of terpenoids. *Nat Biotechnol* 2003;21(7):796-802.
24. Ro DK, Ouellet M, Paradise EM, et al. Induction of multiple pleiotropic drug resistance genes in yeast engineered to produce an increased level of anti-malarial drug precursor, artemisinic acid. *BMC Biotechnol* 2008;8:83.
25. Ro DK, Paradise EM, Ouellet M, et al. Production of the antimalarial drug precursor artemisinic acid in engineered yeast. *Nature* 2006;440(7086):940-943.
26. Paddon CJ, Westfall PJ, Pitera DJ, et al. High-level semi-synthetic production of the potent antimalarial artemisinin. *Nature* 2013;496(7446):528-532.
27. Tsuruta H, Paddon CJ, Eng D, et al. High-level production of amorpha-4,11-diene, a precursor of the antimalarial agent artemisinin, in Escherichia coli. *PLoS One* 2009;4(2):e4489.
28. Westfall PJ, Pitera DJ, Lenihan JR, et al. Production of amorphadiene in yeast, and its conversion to

dihydroartemisinic acid, precursor to the antimalarial agent artemisinin. *Proc Natl Acad Sci U S A* 2012;109(3):E111-118.
29. Spiegel A. It's about time you knew exactly where vanilla comes from. *Huffington Post* March 25, 2014. http://www.huffingtonpost.com/2014/03/25/vanilla-comes-from_n_5021060.html. Accessed August 12, 2016.
30. Watson E. Synthetic biology is cheaper, faster, and more systainable, says Evolva CEO: 'We're proud of what we do'. *Food Navigator* 2014; http://www.foodnavigator-usa.com/Suppliers2/Evolva-on-synthetic-biology-Its-faster-cheaper-more-sustainable. Accessed June 1, 2016.
31. Barclay E. GMOs are old hat. Synthetically modified food is the new frontier. *NPR* October 3, 2014. http://www.npr.org/blogs/thesalt/2014/10/03/353024980/gmos-are-old-hat-synthetically-modified-food-is-the-new-frontier. Accessed August 12, 2016.
32. Code of Federal Regulations. Title 21—Food and Drugs. § 101.222010. https://www.gpo.gov/fdsys/pkg/CFR-2010-title21-vol2/xml/CFR-2010-title21-vol2-sec101-22.xml. Accessed August 12, 2016.
33. Rubber Manufacturers Association. Rubber FAQs. https://rma.org/about-rma/rubber-faqs. Accessed May 20, 2016.
34. Global tire shipments to reach 1.7 billion units by 2015, according to a new report by Global Industry Analysts, Inc. PRWeb May 29, 2016. http://www.prweb.com/releases/tires_OEM/replacement_tire/prweb4545704.htm. Accessed August 12, 2016.
35. Goodyear. Bio-based tires edge closer to reality/ collaboration between the Goodyear Tire & Rubber Company and DuPont Industrial Biosciences results in breakthrough technology for tires made with renewable raw materials [press release]. March 6 2012. https://corporate.goodyear.com/en-US/media/news/Bio-

based-Tires-Edge-Closer-to-Reality—Collaboration-between-The-Goodyear-Tire—Rubber-Company-and-DuPont-Industrial-Biosciences-results-in-breakthrough-technology-for-tires-made-with-renewable-raw-materials-1426100365334.html. Accessed August 12, 2016.
36. Whited GM, Feher FJ, Benko DA, et al. Technology update: development of a gas-phase bioprocess for isoprene-monomer production using metabolic pathway engineering. *Industrial Biotechnology* 2010;6(3):152-163.
37. Erickson B, Singh R, Winters P. Synthetic biology: regulating industry uses of new biotechnologies. *Science* 2011;333(6047):1254-1256.
38. Hayden EC. Deal between Amyris and Michelin highlights industry's hunt for a profitable niche. *Nature* September 30, 2011.
39. Braskem, Amyris and Michelin team up to accelerate development of renewable isoprene. *World Industrial Reporter* September 11, 2014. http://worldindustrialreporter.com/braskem-amyris-michelin-team-accelerate-development-renewable-isoprene/. Accessed August 12, 2016.
40. Goodyear. Innovation through technologies. http://www.goodyear.eu/corporate_emea/our-responsibilities/innovation/through-technologies.jsp. Accessed July 23, 2015.
41. DuPont. SORONA® frequently asked questions. http://www.dupont.com/products-and-services/fabrics-fibers-nonwovens/fibers/brands/dupont-sorona/open/sorona-faq.html. Accessed May 28, 2016.
42. Universal Bio Mining. http://universalbiomining.com/. Accessed July 23, 2015.
43. Römer L, Scheibel T. The elaborate structure of spider silk: structure and function of a natural high performance fiber. *Prion* 2008;2(4):154-161.
44. Bourzac K. Spinning synthetic spider silk. *MIT Technology*

Review September 21, 2015. https://www.technologyreview.com/s/541361/spinning-synthetic-spider-silk/. Accessed August 12, 2016.
45. Rao L. Bolt Threads will bring its spider silk fabric to Patagonia. *Fortune* May 11, 2016. http://fortune.com/2016/05/11/bolt-threads-patagonia/. Accessed August 12, 2016.
46. Algenist. http://www.algenist.com/. 2013. Accessed October 8, 2013.
47. Joint BioEnergy Institute. http://www.jbei.org/. Accessed June 1, 2016.
48. Strickland E. DNA Manufacturing enters the age of mass production. *IEEE Spectrum* December 23, 2015. http://spectrum.ieee.org/biomedical/devices/dna-manufacturing-enters-the-age-of-mass-production. Accessed August 12, 2016.
49. Zhang S. Cheap DNA sequencing is here. Writing DNA is next. *Wired* November 20, 2015. http://www.wired.com/2015/11/making-dna/. Accessed August 12, 2016.
50. Doudna JA, Charpentier E. Genome editing. The new frontier of genome engineering with CRISPR-Cas9. *Science* 2014;346(6213):1258096.
51. Kaminski R, Bella R, Yin C, et al. Excision of HIV-1 DNA by gene editing: a proof-of-concept in vivo study. *Gene Ther* 2016;23(8-9):696.
52. Cressey D, Cyranoski D. Gene editing poses challenges for journals. *Nature* 2015;520(7549):594.
53. Cyranoski D. Ethics of embryo editing divides scientists. *Nature* 2015;519(7543):272.
54. Cyranoski D, Reardon S. Embryo editing sparks epic debate. *Nature* 2015;520(7549):593-594.
55. Adelman ZN, Tu Z. Control of mosquito-borne infectious diseases: sex and gene drive. *Trends Parasitol* 2016;32(3):219-229.
56. Akbari OS, Bellen HJ, Bier E, et al. Biosafety. Safeguarding

gene drive experiments in the laboratory. *Science* 2015;349(6251):927-929.
57. Baltimore D, Berg P, Botchan M, et al. Biotechnology. A prudent path forward for genomic engineering and germline gene modification. *Science* 2015;348(6230):36-38.
58. Gantz VM, Jasinskiene N, Tatarenkova O, et al. Highly efficient Cas9-mediated gene drive for population modification of the malaria vector mosquito Anopheles stephensi. *Proc Natl Acad Sci U S A* 2015;112(49):E6736-6743.
59. Rogers JK, Church GM. Multiplexed engineering in biology. *Trends Biotechnol* 2016;34(3):198-206.
60. Rogers JK, Taylor ND, Church GM. Biosensor-based engineering of biosynthetic pathways. *Curr Opin Biotechnol* 2016;42:84-91.
61. Muscat RA. Synthetic biology comes into its own. *The Scientist* June 1, 2016. http://www.the-scientist.com/?articles.view/articleNo/46170/title/Synthetic-Biology-Comes-into-Its-Own/#ref. Accessed August 12, 2016.
62. Bright P. Microsoft experiments with DNA storage: 1,000,000,000 TB in a gram. *Ars Technica* April 27, 2016. http://arstechnica.com/information-technology/2016/04/microsoft-experiments-with-dna-storage-1000000000-tb-in-a-gram/. Accessed August 12, 2016.
63. Patel P. Tech turns to biology as data storage needs explode. *Scientific American* May 31, 2016. http://www.scientificamerican.com/article/tech-turns-to-biology-as-data-storage-needs-explode/. Accessed August 12, 2016.
64. Yin S. Is DIY kitchen CRISPR a class issue? *Popular Science* May 3, 2016. http://www.popsci.com/is-bringing-crispr-to-kitchens-class-issue. Accessed August 12, 2016.
65. Synbiota: About us. https://synbiota.com/about. Accessed July 20, 2015.

66. Wohlsen M. *Biopunk Solving Biotech's Biggest Problems in Kitchens and Garages*. New York: Current; 2011.
67. Schwartz J. Sushi study finds deception. *New York Times* August 22, 2008. http://www.nytimes.com/2008/08/22/world/americas/22iht-fish.1.15539112.html?_r=0. Accessed August 12, 2016.
68. Peckham M. Texas apartment to track dog poop offenders using DNA. *Time* 2013. http://newsfeed.time.com/2013/01/24/texas-apartment-to-track-dog-poop-offenders-using-dna/. Accessed August 12, 2016.
69. Yetisen AK, Davis J, Coskun AF, Church GM, Hyun Yun S. Bioart. *Trends Biotechnol* 2015;33(12):724-734.
70. Genspace. Invasion ecology: turning weeds into watercolors. 2016. http://bit.ly/2dNEJGv. Accessed August 12, 2016.
71. Johnson RA. Synthetic biology: 10 policy reasons it matters to U.S. foreign policy. National Academy of Sciences, Board on Life Sciences, NAS Forum on Synthetic Biology. The Evolving Nature of Synthetic Biology: A Panel Discussion on Key Science, Policy, and Societal Challenges Facing the International Community. Washington, DC: State Department; 2013. http://sites.nationalacademies.org/cs/groups/pgasite/documents/webpage/pga_084543.pdf. Accessed August 12, 2016.
72. Florea M, Hagemann H, Santosa G, et al. Engineering control of bacterial cellulose production using a genetic toolkit and a new cellulose-producing strain. *Proc Natl Acad Sci U S A* 2016;113(24):E3431-E3440.
73. Smith C. Engineering 'tea bacteria' could lead to advanced materials. Imperial College London *News* May 30, 2016. http://bit.ly/2dNEdZ1. Accessed August 12, 2016.
74. iGEM. Synthetic biology based on standard parts. http://igem.org/About. Accessed April 1, 2016.
75. Interview with Randy Rettberg by Gigi Kwik Gronvall (2013).

76. DIY Bio. Local groups. http://diybio.org/local/. Accessed June 2, 2016.
77. Jozuka E. This biohacker BBQ would be way cooler with less red tape. *Motherboard* April 21, 2015. http://motherboard.vice.com/en_uk/read/this-biohacker-bbq-would-be-way-cooler-with-less-red-tape. Accessed July 21, 2015.
78. Carlson R. *Biodefense Net Assessment: Causes and Consequences of Bioeconomic Proliferation: Implications for U.S. Physical and Economic Security.* Department of Homeland Security Science and Technology Directorate;2012. http://bit.ly/2dmNWK3
79. Grush L. SXSW 2015: I reprogrammed a lifeform in someone's kitchen while drinking a beer. *Popular Science* March 14, 2015. http://www.popsci.com/sxsw-2015-i-made-recombinant-dna-someones-kitchen-while-drinking-beer. Accessed August 12, 2016.
80. BosLab. Truffle hacking project. http://www.boslab.org/#!truffle-hacking-project/pmzto. Accessed May 20, 2016.
81. CBS News. Truffles: the most expensive food in the world. *60 Minutes* June 4, 2012. http://www.cbsnews.com/news/truffles-the-most-expensive-food-in-the-world/. Accessed August 12, 2016.
82. Mole B. Microbes' role in truffle scents not trifling. *ScienceNews* July 27, 2015. https://www.sciencenews.org/article/microbes%E2%80%99-role-truffle-scents-not-trifling. Accessed August 12, 2016.
83. Vahdatzadeh M, Deveau A, Splivallo R. The role of the microbiome of truffles in aroma formation: a meta-analysis approach. *Appl Environ Microbiol* 2015;81(20):6946-6952.
84. Real Vegan Cheese! Indiegogo. 2016. https://www.indiegogo.com/projects/real-vegan-cheese/#/. Accessed August 12, 2016.
85. Hayden EC. Bioengineers debate use of military money.

Nature 2011;479(458). http://www.nature.com/news/bioengineers-debate-use-of-military-money-1.9409. Accessed August 12, 2016.
86. Office of Technical Intelligence, Office of the Assistant Secretary of Defense for Research and Engineering. *Technical Assessment: Synthetic Biology.* Department of Defense Research and Engineering. January 2015. http://defenseinnovationmarketplace.mil/resources/OTI-SyntheticBiologyTechnicalAssessment.pdf. Accessed August 12, 2016.
87. Tucker P. Big data, synthetic biology, and space planes are the weapons of the future. *Defense One* March 26, 2014. http://www.defenseone.com/technology/2014/03/big-data-synthetic-biology-and-space-planes-are-weapons-future/81346/. Accessed August 12, 2016.
88. Rambhia KJ, Ribner AS, Gronvall GK. Everywhere you look: select agent pathogens. *Biosecur Bioterror* 2011;9(1):69-71.
89. World Health Organization. *Global Eradication of Smallpox: Final Report of the Global Commission for the Certification of Smallpox Eradication, December 1979.* Geneva. http://whqlibdoc.who.int/publications/a41438.pdf. Accessed August 12, 2016.
90. Church G. De-extinction is a good idea. *Scientific American* September 1, 2013. http://www.scientificamerican.com/article.cfm?id=george-church-de-extinction-is-a-good-idea. Accessed August 12, 2016.
91. University of Glasgow. Frontier engineering—synthetic biology applications to water supply and remediation. http://wateratglasgow.org/frontier-engineering-synthetic-biology-applications-to-water-supply-and-remediation/. Accessed August 12, 2016.
92. Gronvall GK. H5N1: a case study for dual-use research. Council on Foreign Relations Working Paper. July 2013. http://www.cfr.org/public-health-threats-and-pandemics/

h5n1-case-study-dual-use-research/p30711. Accessed August 12, 2016.

93. Kelley NJ. The promise and challenge of engineering biology in the United States. *Industrial Biotechnology* 2014;10(3):137-139.
94. The White House. *National Security Strategy.* February 2015. https://www.whitehouse.gov/sites/default/files/docs/2015_national_security_strategy.pdf. Accessed August 12, 2016.
95. Bergin J. *Synthetic Biology: Global Markets.* BCC Research. June 2014. http://www.bccresearch.com/market-research/biotechnology/synthetic-biology-bio066c.html. Accessed August 12, 2016.

CHAPTER 2.

ON SECURITY

Given the devastation that naturally occurring infectious diseases can inflict—the Ebola crisis in West Africa in 2014-15 killed more than 11,000 people, devastated the already strained economies of Guinea, Sierra Leone, and Liberia, and cost the US government $5.5 billion in its response—the potential for a biological weapons attack on an unsuspecting population may sound more like a movie than real life.[1-3] In fact, there are health leaders who often remark that "Nature is the worst bioterrorist."[4-6] MERS, chikungunya, and Zika virus add to the already long list of natural biological threats for which there are inadequate vaccines, drugs, or other medical countermeasures. We are not prepared for many current diseases that threaten us, and there is always another one that we don't know about yet, just over the horizon. The threat of natural disease deserves a great deal of our medical response and preparedness resources.

But the possibility of a thinking enemy needs to be prepared for, too. Deliberate use of biological agents for harm is thankfully not an everyday event, but it does happen. Throughout history, biological capabilities have been applied to weapons and warfare. Many nations, including the United States, also had biological weapons programs until a decision was made in 1969 to give them up. The Convention on the Prohibition of the Development, Production and Stockpiling of Bacteriological

(Biological) and Toxin Weapons and on their Destruction (referred to as the Biological Weapons Convention, or BWC) was negotiated, which banned biological weapons; it was signed in 1972 and ratified in the United States in 1975.[7] Since the BWC has been in effect, the majority of nations have ended any offensive biological weapons development or use. However, there have been multiple documented violations, including by South Africa, by Iraq before the first Gulf War, and, most egregiously, by the former Soviet Union. There have been bioterrorist incidents and attempts, from the Rajneesh in Oregon in 1984, in which cult members poisoned salad bars with salmonella to influence a local election, sickening 751 people, to the Aum Shinrikyo cult, in which members tried to release anthrax into the air multiple times before successfully mounting a chemical weapons attack in a Tokyo subway in 1995, to the anthrax letter attacks in 2001.[8,9] More recently, there have been criminal acts with biological agents and toxins, particularly ricin.[10,11] And various violent nonstate actor groups, including Islamic State of Iraq and Syria (ISIS) and Al Qaeda, are thought be seeking biological weapons capabilities.[12-14]

The long history of attempts to develop and use biological weapons does not bode well for the future, but the outlook is even more alarming if one considers how current biological techniques could be applied to a weapon. There has been a great deal of scientific progress since the United States shut down its weapons program in 1969 and since the anthrax letter attacks in 2001. Rather than nature being the "worst bioterrorist," a biological weapon could be deliberately designed that is as bad as or worse than what nature could produce. Also, a thinking enemy could devise methods to deliver a biological weapon that are not typical of a natural pathogen. The possibility that new biotechnologies, including from the relatively new and promising field of synthetic biology, could aid biological weapons development has alarmed security experts and policymakers.

Synthetic biology is a scientific discipline that aims to make biology easier to engineer, and the field has produced a number of powerful tools that make biological systems easier to manipulate. It has already delivered on the faster production of influenza vaccines, an antimalarial drug, and new food products, such as a vanilla flavoring that does not have to be extracted from vanilla beans.[15] The tools developed in synthetic biology have spread beyond the discipline into all of biological research and biotechnology: for example, the ability to perform gene synthesis, with accuracy, speed, and at low costs, and CRISPR/Cas9, a gene-editing tool that allows sections of DNA to be cut and pasted as in a Word document. These technologies are boons for biological research, commercial development, and many beneficial medical applications. But if these powerful technologies and others were to be misapplied to the development of a biological weapon, it could help to make already dangerous pathogens worse, by making them more difficult to detect, prevent, or treat. It is the possibility for misuse to cause harm that drove gene-editing technologies to be added to the annual worldwide threat assessment report of the US intelligence community in 2016.[16,17]

While new technologies are inevitably a concern for biodefense, the simple, unfortunate truth is that the development of biological weapons does not require synthetic biology or new biotechnologies. The technologies that were available to defunct biological weapons programs in the 1960s are still available now—but the laboratory methods used in those old programs are more accessible and less costly. The basic starting materials for weaponization of biology—pathogens that affect humans—are found in nature, laboratories, and sick people all over the world.[18] Even the *Bacillus anthracis* strain that was used in the anthrax letter attacks of 2001 was not a sophisticated laboratory or synthetic creation, but an ordinary bacteria isolated from an infected cow in Texas decades earlier.[19] The technologies and methods to cultivate and weaponize a variety of pathogens are widely accessible because they have considerable overlap with nonbioweapons methods and technologies that are actively pursued for beneficial purposes. In other words, these technologies are dual use and are pursued for important and legitimate purposes.

Even though there are many less-sophisticated methods that could be used to make a biological weapon, the possibility that synthetic biology could be used for weapons development should be a concern for several reasons: first, because synthetic biology could be used to increase the types and accessibility of biological weapons that are not available using more conventional methods, and second, because synthetic biology lowers the barriers to biological weapons by making biological tools more accessible and powerful. Using synthetic biology, it would not be necessary to isolate a pathogen from an environmental sample or from a sick patient before developing it as a weapon. The ability to recreate a pathogen without this harvesting step, and to make it "from scratch," could also allow the weaponization of eradicated or difficult-to-access pathogens.

Bringing Back Smallpox

One such pathogen is smallpox, which has been eradicated from the natural world. Smallpox had a 30% mortality rate and killed more people in the 20th century—the century in which it was eradicated—than WWI, WWII, and all other wars and armed conflicts in the century *combined*.[20] After a heroic effort to eradicate smallpox in the natural world, which was declared accomplished in 1980, a strict international regulatory regime coordinated by the World Health Organization (WHO) was developed. Each and every experiment on the virus is controlled, and the intact virus is allowed to be present in only 2 facilities in the world: a laboratory now located in the Russian Federation and in the Centers for Disease Control and Prevention in Atlanta, Georgia.

Genetic sequencing of smallpox was performed when it became technically feasible.[21] The virus was sequenced to gain understanding of how smallpox was able to defeat the human immune system and for other medically important reasons, including determining whether other pox viruses in the world were likely to mutate into a smallpox-like virus. For example, a cousin to smallpox, monkeypox, is able to infect humans, and it sometimes spreads from person to person, though it does not spread as rapidly as smallpox. It was an open question whether monkeypox might mutate to become more virulent and fill the niche left by smallpox after it was eradicated.[22-24] For years after smallpox was eradicated, the World Health Organization surveilled for monkeypox infections in Africa. More recently, it is becoming clear that monkeypox is resurging in West Africa, with greater numbers of human cases, transmitted from monkeys and from squirrels.[25]

In 1994, all of the A's, G's, C's, and T's that make up the complete genetic sequence of the smallpox virus were published, and now they are freely available on the internet.[21] When the sequencing work commenced, technologies for "reading" DNA (DNA sequencing) were much more advanced than those for "writing" DNA (gene synthesis).[21] At that time, it was not even remotely possible to synthesize a genome the size of smallpox, a large virus that is over 186,000 base pairs of DNA. In fact, the assembly of 30 base pairs would have been a challenge. However, the ability to write DNA caught up rapidly with advances in technology. When smallpox was originally sequenced, nobody anticipated that, as a 2010 World Health Organization report put it, "advances in genome sequencing and gene synthesis would render substantial portions of [smallpox virus] accessible to anyone with an internet connection and access to a DNA synthesizer."[26(p45)]

Making an infectious smallpox virus from its genetic sequence is a complicated endeavor that may only be possible by highly sophisticated laboratories at this time, but it is certain that it will continue to get easier for many others to perform. It is now as hard to recreate the smallpox virus as it will ever get—certainly, advances in synthetic biology and other related technologies will not make the process *more* difficult. Routine vaccination of children ceased in 1972, as did the booster shots that were required every 10 years for complete protection. Virtually everyone on the planet is thus vulnerable to smallpox should it appear again. The United States, Singapore, and several other countries assessed that the risks of the return of smallpox as a weapon were too great and that medical countermeasures to treat and immunize against smallpox should be stockpiled and should be maintained as a deterrent to attack.[27]

Smallpox and rinderpest, a disease that mostly affected cattle, are the only diseases that have been eradicated from the planet. But even some viruses that are still able to be found in the natural world may be easier to acquire through synthesis in the laboratory than from the environment or from laboratory samples. For example, isolating the Ebola virus or foot and mouth disease (FMD) viruses from their natural environments or from a laboratory may require nefarious actors to take steps that could compromise their operational security, require more skill than they possess to harvest the samples from nature, or introduce much more uncertainty into reaching their goal. Regulations and controls on the access to laboratory stocks of pathogens make acquisition from laboratories much more difficult as well as being traceable; regulations on pathogen access used to be concentrated in the United States, but access controls are increasingly common in other nations in the world as well. These programs cannot remove the threat of misuse entirely—and there is a point at which those security measures have diminishing returns and interfere with beneficial research—but it is a sound strategy to continue to raise barriers to easy access to pathogens.

A synthesized, laboratory-made copy of a known dangerous pathogen, or a slightly modified version, made with easily accessible technologies and materials, is a very serious problem for biodefense. The tools available to prevent such a scenario are limited. To boost security against these threats, it is important to detect threats early and to apply defenses—medical and public health interventions, as well as attribution—as quickly as possible to mitigate loss of life. It is possible that a very skilled actor or well-resourced group could use synthetic biology to develop a more novel pathogen (eg, a variant of influenza to which a population does not have immunity) or to modify pathogens so that stockpiled medical countermeasures, diagnostic tests, and detectors are not effective. This could lead to delays in diagnosing patients and implementing public health interventions.

Controls on DNA Synthesis

Even before 2001, when concerns about bioterrorism became more prevalent in the United States, practitioners of the then-tiny field of synthetic biology were conscious that synthetic biology could be applied to weapons development and that synthetic biology could add to the already existing problems of biological weapons and biological safety.[28] In the First International Conference on Synthetic Biology 1.0 in 2004 (as well as subsequent conferences titled SB 2.0, 3.0, etc), efforts to mitigate the risks were discussed, in such topics as gene synthesis screening, codes of conduct, normative ethics, and obligations to report misuse of laboratory equipment or skills.[28,29]

The synthetic biology community's willing early engagement in ethical discussions and their proposals for limits to scientific investigation for the greater good was encouraging, leading philanthropic foundations and the Federal Bureau of Investigation (FBI) to fund the field's progress in self-governance. The Alfred P. Sloan Foundation developed a program area to identify and address the societal, ethical, and regulatory risks associated with synthetic biology and its applications to ensure that the field could advance in the public interest. Many of the tangible policy advances toward these goals were fueled by Sloan's investments, particularly the Synthetic Biology Project at the Woodrow Wilson International Center for Scholars.[30] The FBI has been proactive in addressing the risks of synthetic biology, with a successful outreach program to university-based as well as amateur biologists, to raise their awareness that their scientific work could be misused for harm and to give them points of contact to report suspicious behavior. These awareness-raising mechanisms may forestall an incident and give law enforcement a chance to intervene to stop malicious actors.[31,32]

The security risks of synthetic biology were known to a community of experts and practitioners, but, in general, policymakers and the public were not aware of the biosecurity implications—until the publication of a 2002 *Science* magazine article that demonstrated the creation of poliovirus using mail-ordered DNA segments.[33] The researchers described how they chemically strung together 60 base pair pieces of DNA to build DNA molecules encoding the ~7,500 base-pair virus. Once the DNA was transcribed into RNA and translated into proteins in a test tube, the researchers had infectious polio virus. The publication caused additional alarm among policymakers when it became widely known that the genetic information for polio virus, as with most other viruses, is available on the internet. With gene synthesis technology accessible to people all over the world, there would be little to prevent a nefarious actor from making pathogens from scratch, including the eradicated smallpox virus.

The advances in gene synthesis have been rapid and surprising and are certain to continue. It took 3 years to perform the polio work, but 2 years after the polio work was published, it took just 2 weeks for the J. Craig Venter Institute (JCVI) to synthesize a similarly sized bacteriophage, which is a virus that infects bacteria. This demonstrated that the technical barriers to making organisms from their genetic information were rapidly dropping. In 2005, the 1918 Spanish influenza virus—responsible for the deaths of more than 100 million people—was reconstructed.[34] In 2010, Craig Venter and colleagues synthesized an entire bacterial genome and "booted up" a synthetic cell—an organism that is much larger and more complicated than any virus, including smallpox.[35] They followed up this feat in 2016 with the 3.0 version of their synthetic organism, with a pared-down genome that nonetheless has all the genes essential for life.[36]

The limitation of pathogen-based security controls in the advent of synthetic biology became further evident to the public after a reporter from *The Guardian*, a UK-based newspaper, placed an online order to a gene-synthesis biotechnology company for a piece of smallpox DNA. The order was to be delivered to a residential address, using a made-up company name—and the DNA was successfully procured by the reporter.[37] The complete smallpox virus is 52,000 base pairs, much larger than the 78-base pair piece the reporter ordered, and there was no danger to the public, but the point was made: These gene synthesis companies could unwittingly provide services for nefarious actors. When the story was revealed, the synthesis company admitted that they did not screen their orders to detect people ordering the DNA of pathogens, and there were no security checks on who could place the orders.[37]

Scientists at the time decried the reporter's efforts as an unnecessarily inflammatory stunt and criticized the *Guardian* article for downplaying the technical expertise required to assemble the complete smallpox virus. The regulatory problem to ban synthesis of smallpox is also technically complex, as there is a 97% similarity between smallpox and another virus, vaccinia. Vaccinia is the virus used to vaccinate against smallpox, and it is also used in other medical research. Given this similarity, it is difficult to draw a regulatory red line about what DNA sequences are "dangerous."[38,39] These technical objections were overruled, however. The newspaper report, along with the drumbeat of new, alarming demonstrations of scientific advances in synthetic biology, drew political attention to the new security risks inherent in emerging biotechnologies. It became clear that many of the regulations intended to prevent bioterrorism, which relied on controlling physical access to pathogens, could be defeated through synthetic biology.

Steps to make synthetic biology more secure were taken. The Alfred P. Sloan Foundation funded the creation of the first sequence screening software, which a gene synthesis company could use to alert them to suspicious orders.[40] Some synthesis companies banded together to form codes of conduct and to agree to screen their orders.[41] The synthetic biology community instituted pledges to avoid using synthesis companies that do not perform screening.[42] In 2010, the US Department of Health and Human Services (HHS) released guidance to petition suppliers of double-stranded DNA to screen their orders, looking for prohibited pathogen sequence matches.[43] The HHS guidance also called for enhanced customer screening, to ensure compliance with US trade restrictions and export controls. If sequence screening determines that a customer has requested genetic material available only to those with clearance to work with select agents, those pathogens that are regulated in the

United States, then the customer must be in compliance with select agent regulations.

Some analysts hoped that the US government would impose stricter controls on gene synthesis, but more stringent regulations would be difficult to put into practice.[44,45] Gene synthesis is an international business, and companies outside the United States are not subject to US regulations. If it were too onerous to go through commercial suppliers, most people would make the genes they need for their research themselves, obviating the usefulness of regulating commercial suppliers.

Not all international gene synthesis companies are members of an industry organization that agrees to either customer screening or sequence screening; thus, there are opportunities for the United States to encourage other nations to issue similar guidance, promote industry-wide screening standards, and champion a common code of conduct for suppliers of DNA. The US should reach out to other nations, especially China, which has numerous gene synthesis companies, to follow guidance similar to the 2010 US HHS recommendations for sequence screening. The United States should work to increase the number of companies that perform gene screening, promote awareness of companies' codes of conduct, and encourage providing those companies with a means to report a suspicious order. The BWC meetings, held twice a year in Geneva, Switzerland, could be one of the forums at which the United States works with other countries to enact similar screening measures.[46]

Though the screening of gene synthesis orders should be expanded, this will not eliminate the risks. The gene synthesis companies typically perform services more cheaply and with greater reliability than would be possible for most laboratories to achieve on their own, but they are not strictly necessary. A bad actor intent on bioterrorism could acquire the starting materials to synthesize genes on his or her own. Like other biotechnologies, the capability for individuals to synthesize genes or whole viruses is increasing, while costs are dropping. There are also signs that the screening technologies are becoming less appealing to companies from a purely business sense: The screening procedures are a stable cost and require a significant expenditure of time. Meanwhile, the costs to synthesize genes are dropping and taking less time. The security screening of DNA orders is thus becoming a more substantial part of a commercial DNA synthesis company's operating costs.[47] The US Intelligence Advanced Research Projects Activity (IARPA) is hoping to automate this screening, funding a research program to develop better approaches to characterizing gene functions based on genetic sequences.[48]

Importantly, there is no publicly available data about how valuable the sequence screening actually is in stopping misuse. It would be useful to know just how often suspicious orders are flagged based on the sequence alone, compared to customer screening information such as methods of payment and address of the customer. It could be that customer screening, and not sequence screening, is the most valuable part of the security screening process. The United States should fund a pilot study to determine the costs and benefits of screening DNA orders. This data could help to shape guidance and regulations so that they are effective as well as attractive for companies.

The Uphill Battle to Develop Biodefenses

As a class, biological weapons pose a particularly challenging national security problem because they could be highly lethal, hard to detect, and difficult to attribute. They are hard to prioritize and expensive to prepare for even without the addition of synthetic biology: There are many different types of pathogens that could be weaponized, which all require separate mechanisms for prevention or response. A vaccine for anthrax is not the same as a vaccine for smallpox. The medical response and public health interventions that would be required would be different, shaped by the specifics of the pathogen and how it infects and spreads. A biological weapon, once released, will reduce the perpetrator's stockpile, but only temporarily: In contrast to other types of weapons, once a bad actor has the capability to produce the weapon, he or she may grow more using the same processes used the first time, because living organisms reproduce and grow. The ability to keep on attacking without a depletion of a biological weapons stockpile has been termed the "reload" effect.[49]

Intelligence about bioweapons programs and groups' intent to use them has been historically difficult to acquire, and conclusions from intelligence have often been drawn in error. There were few indications during the Cold War that the Soviet Union employed tens of thousands of scientists dedicated to making biological weapons (including smallpox, Ebola, and anthrax) or that Iraq was developing biological weapons at the time of the first Gulf War. Actual and attempted biological attacks, including those perpetrated by the Rajneesh cult and Aum Shinrikyo, were not discovered until well after the fact. Conclusions from intelligence have resulted in both type 1 errors (Iraq was thought to have a biological weapons program during the lead-up to the second Gulf War, when it did not) and type 2 errors (the former Soviet Union was not thought to have a biological weapons program, until defectors revealed the expansive program). Even when information is acquired that points to a specific interest in developing biological weapons, there is still a great deal of uncertainty about whether this amounts to a committed or capable effort or if efforts are already under way. For example, a laptop belonging to a member of ISIS was found to contain instructions in Arabic for weaponizing bubonic plague isolated from infected animals.[13] The true operational threat posed by this information remains unknown.

Microbial forensics can be used to determine the source of a microbe, which is important not only for attribution of a biological attack, but also for deterring those who might become interested in using the weapons.[50] Advances have been made in this field, even since a specific strain of anthrax found in a US army laboratory was tied to anthrax included in letters sent to several congressional offices in 2001.[51] Still, there are many conceivable scenarios in which it would be difficult to attribute a biological attack or even definitively declare that a disease outbreak is the result of an attack. This ambiguity in attributing the crime, as well as the multiple skills required to successfully perpetrate a biological weapons attack, could lead to biological attacks for commercial gain. For example, a deliberate outbreak of a crop-killing fungus, or the introduction of FMD in a nation that formerly held FMD-free status, would harm that country's economy. Billions of dollars would need to be spent on culling infected and exposed animals, as well as on vaccination programs, and there would be a loss of revenue for red meat exports until FMD-free status could be regained. On a subnational level, companies could have their fortunes rise or fall depending on outbreaks of food poisoning.

A more extreme scenario, with an entirely novel pathogen developed using synthetic biology, has been a concern of policymakers and has occasionally captured the attention of the media. This scenario also inspired an influential article in the *Atlantic* magazine about a pathogen designed to target just 1 person—the US president.[52] Individually tailored technologies for assassination are theoretically possible in the future, building on ongoing research into developing individually tailored cancer therapies. The assassination threat was described as "not so far-fetched," according to synthetic biologist and Harvard professor George Church in an interview about the *Atlantic* article in

2011.[53] Still, such an endeavor would not be easy or likely for the immediate future, because of the amount of research funding and time required. Even in a post-CRISPR world, there are numerous technical challenges involved in such an effort, because such an individually tailored weapon would require a great deal of research, development, and testing to bring into reality, and just like many other scientific projects, they are likely to fail many times over before they succeed. Testing and refining would be complicated for a designed pathogen that has only 1 lethal target. So-called "ethnic" weapons, a longtime worry about future biowarfare scenarios, would be similarly tricky to research, design, and test and are likely to have considerable overlap with nontarget populations, in spite of a growing body of research that demonstrates that there are some differences in immunity based on ancestry.[54]

But though it is possible that the scientific challenges to such a bioweapon may eventually be surmountable, there are other considerations beyond the technical: As futurist Jamais Cascio put it, in response to the *Atlantic* article, "It's not the science, but the implementation."[53] A plot to accomplish such a specific targeting feat would also run the risk of being exposed before it could be brought to fruition, especially given the number of people who would be required to develop, test, and deliver the weapon and the timelines involved in its development. Delivery of the modified weapon—in the *Atlantic* article scenario, the assassination virus would spread naturally among a group to its intended target—would also not be guaranteed and is of uncertain statistical likelihood. The likelihood could be lowered further if people would adhere to CDC-recommended hygiene etiquette and cover their mouths when they cough and sneeze.[55]

Given the complexity of an individual assassination effort, concerns about an entirely novel bioweapon that is not heavily based on an existing disease are now outside of the realm of a non-state actor group or an individual bad actor. Perhaps they are in the realm of nation states that can afford to commit large amounts of resources to this endeavor. But at this time, there are certainly—and unfortunately—many easier paths to "success" with a simpler, less novel biological weapon. There are many other reasons besides a targeted bioweapon for the Secret Service to try to limit access to the president's coffee cups, or places where the president's DNA could be found: The detection of a cancer gene or Alzheimer's gene, for example, could prove to be embarrassing or politically manipulable. The collection of world leaders' DNA that the US is reported to have is not surprising in this context; having access to genomic information could help to understand travel history, family relationships, and predilections to disease. Other government-funded tools have been used to acquire similar information about world leaders.[52] As an example, an *Atlantic* article from 2005 describes the DoD-funded "movement analysis" that suggested that Vladimir Putin, the leader of the Russian Federation, may have suffered a stroke before he was born.[56]

Deterrence Measures for Biological Weapons of All Types

There are a number of controls already in place intended to deter nations, groups, or individuals from pursuing biological weapons, regardless of whether the weapons were made using standard microbiological methods or through advanced synthetic technologies.

The US participates in international agreements that prohibit biological weapons development and use, especially the BWC. Signed in 1972, it was the first agreement among nations that ruled an entire category of weapons off-limits.[7] There are at present 174 states parties and 8 signatories that have agreed "never in any circumstance to develop, produce, stockpile or otherwise acquire or retain: Microbial or other biological agents, or toxins whatever their origin or method of production, of types and in quantities that have no justification for prophylactic, protective or other peaceful purposes; weapons, equipment or means of delivery designed to use such agents or toxins for hostile purposes or in armed conflict."[7]

Even though there are almost certainly signatories to the treaty that have a clandestine biological weapons program, no nation claims to have a biological weapons program in violation of the treaty or has public biological weapons ambitions. Critics point out that the BWC does not have a verification mechanism to ensure that states are in compliance and that the treaty has been violated several times in the past, most notably by the Soviet Union. However, the fact that materials, equipment, and technical processes for legitimate work can also be directly applicable to weapons development makes a verification mechanism unfeasible. The BWC has been valuable, because it has reinforced the norm against biological weapons and served as a vehicle to discuss other issues, such as protections against biological accidents or the security vulnerabilities that go along with synthetic biology.

In addition to the BWC, UN Resolution 1540 is another legally binding mechanism that is intended to encourage nations to fight terrorism by subnational groups and requires nations to have and enforce measures against the proliferation of nuclear, chemical, and biological weapons.[57]

Pathogen access is regulated in the United States, as well: Laboratories that house a group of pathogens known as "select agents" must be inspected by the CDC or the USDA, and personnel with access require background investigations performed by the Department of Justice.[58] Even if a pathogen is not on the list, the possession of a biological agent, toxin, or delivery system that is not reasonably justified by prophylactic, protective, bona fide research or other peaceful purpose would be in violation of the USA Patriot Act.[59] Many other nations have similar legislation or are engaged in adopting it, a process that is actively encouraged by the US government. There are limitations to a pathogen-based regulatory system, however, as pathogens are found nearly everywhere and can be harvested from other nonlaboratory sources. As stated previously, synthetic biology also will make this a less-effective mechanism, if pathogens can be made through chemical synthesis.

The Control of Biological Information

The most difficult part of bioterrorism prevention is what to do with the life sciences information and data that could lower barriers to biological weapons development or use. The biological sciences are inherently dual use, so a great deal of the scientific knowledge, materials, and techniques required for legitimate, beneficent biological research could be misused to make a biological weapon. Laboratory research conducted to uncover critical information about how a pathogen manipulates the human immune system to cause disease could be exploited to make a disease harder to treat. The scientific community relies on unfettered access to publications, genetic sequences, and biological materials to advance science and, importantly, to reproduce the results of others to verify gains and build on them, so the challenge of protecting information and allowing it to be used for research is a "dual-use dilemma."

The dual-use dilemma has been debated by scientists, ethicists, and policymakers multiple times in the past 15 years, often precipitated by a particularly problematic scientific paper that drew concern and shaped the arguments over what should be done. In 2001, one such paper, published by the *Journal of Virology,* led to an important security implication: that stockpiling smallpox vaccine might not be sufficient in the event of a smallpox attack and that medical therapeutics might be necessary as well.[60] The reason the work was funded was because the Australian government wanted biological control measures to reduce the yearly mouse plagues that destroy crop lands and rural infrastructure. The scientists found that by adding a single gene that modulates immune response to the mousepox virus (named interleukin-4, or IL-4), the modified virus was lethal even to mice that were vaccinated against mousepox. Mousepox is a distant cousin to the human smallpox virus, so this research raised fears that the same genetic modification could be repeated with smallpox virus, so that a smallpox with added IL-4 would be lethal, even to those who had been vaccinated against unmodified smallpox.

The National Academies of Science, funded by the Alfred P. Sloan Foundation, took this problem on to make recommendations for how the scientific community should deal with dual-use research like the mousepox case.[61] Their findings held increased weight in the aftermath of the attacks on September 11, 2001, and the subsequent anthrax letter attacks. The report, *Biotechnology in an Age of Terrorism,* also known as the "Fink report" after its chairman, MIT geneticist Dr. Gerald R. Fink, made the case that scientists have an "affirmative moral duty to avoid contributing to the advancement of biowarfare or bioterrorism."[61(p4)] The report recommended that scientists give extra consideration and review before undertaking projects that have been nicknamed the "7 deadly sins": projects that demonstrate how to make a vaccine ineffective; confer resistance to therapeutically useful antibiotics or antiviral agents; enhance the virulence of a pathogen or render a nonpathogen virulent; increase a pathogen's transmissibility; alter the host range of a pathogen; enable the evasion of a diagnostic and/or detection modalities; or enable the weaponization of a biological agent or toxin.[61]

Though the weaponization of *any* biological agent is clearly forbidden by law and by treaty, it is worth noting that the other "experiments of concern" listed are not ruled off-limits, but should require extra review as the work, even if legitimate, is more likely to have the potential to be misused. This is important, because there are good scientific reasons for pursuing for beneficial purposes experiments that would fall into these categories. For example, the host range of a pathogen might be expanded in order to create an animal model to further study a human disease. Resistance to a new antibiotic could be cultivated in the laboratory to determine whether and by what pathway resistance will be acquired in the real world. These categories may sound alarming, but it is the context of the scientific work and the intent of the scientists performing it that change how the work is perceived.

The Fink report's 7 experiments of concern became the starting point for an HHS advisory committee, the National Science Advisory Board for Biosecurity (NSABB), formed as a recommendation of the Fink report in 2004 to consider the dual-use issue. The NSABB codified dual-use research of concern (known as DURC) as "life sciences research that, based on current understanding, can be reasonably anticipated to provide knowledge, information, products, or technologies that could be directly misapplied to pose a significant threat with broad potential consequences to public health and safety, agricultural crops and other plants, animals, the environment, materiel, or national security."[62(pp1-2)] DURC review is now required for US federally funded research with regulated pathogens, and scientists are required to develop a risk mitigation plan and assess risks and benefits of the research.[62]

Acting on DURC

Defining DURC is easier than knowing what to do about it.

Over the past 15 years, numerous legitimate and informative biological studies have fallen into the DURC category, raising security questions and splitting the opinions of scientists, ethicists, and policymakers on whether the research should have been performed or published. Often, the divisive research has involved non-select agents, pathogens that are not subject to security regulations, so they would not be covered by DURC review.

Part of the difficulty in overseeing DURC is that control measures would be highly dependent on how likely the information is to be misused. It is hard to know, without extraordinary insight into the minds and plans of would-be bioterrorists, just how useful a specific scientific insight or series of papers is likely to be. In addition, there are multiple actors, with different levels of skill, who might misuse the biological research. Without a defined malevolent actor in mind who might misuse the information to create a biological weapon—one who has the necessary skills to manipulate biological systems as well as resources—it is a challenging process to balance the risks of publication and subsequent misuse along with the potential for a more positive outcome. On one end of the spectrum there may be biological science amateurs who could be looking for step-by-step guidance from the internet, who are not likely to be able to incorporate new insights into how to make an already serious disease worse. Though there are methods sections in scientific papers that exist to enable replication of the scientific study, these sections are not nearly as descriptive as a recipe in a cookbook. A great deal of knowledge is assumed for those who are skilled in the field, and the methods are not explicitly explained—this has been called "tacit knowledge."[63] On the other end of the spectrum, there may be state-sponsored scientists with the time and resources to perform R&D and who are looking at the scientific literature for weaponizable insights. Given the incremental nature of science, it is also not clear how one paper among many could affect their pursuits or significantly lower their barriers to biological weapons development.

There is also an abundance of dual-use research that could give insights into the development of many different types of biological weapons; well-resourced groups have an almost limitless array of technical options they could pursue for harm. In addition to the examples already discussed, there are many well-known demonstrations of legitimate scientific work that could be misused in order to make pathogens more effective as biological weapons, including making antibiotic-resistant bacteria,[64,65] engineering viruses that can escape their vaccines,[60] engineering anthrax bacteria so that the vaccine is ineffective,[66] and making viruses more transmissible.[67,68] In addition, there have been advances in neuroscience and its effects on behavior that could be misused,[69] as could the field of paleovirology, in which scientists have been able to resurrect extinct retroviruses that are millions of years old and then demonstrate their ability to infect human cells *in vitro*.[70]

Projecting changes in technologies that increase threats is also open to error. The historical example of the sequencing of the smallpox virus demonstrates this point. If all stocks of the smallpox virus were destroyed in 1980, without ever being sequenced, information valuable to infectious disease research would have been lost forever. Yet, now that the sequence is known, it is possible that smallpox could be remade. The likelihood of smallpox's reintroduction to nature is unknown—and perhaps unknowable—while the positive contributions of the sequencing of the smallpox virus and subsequent research are immense and measurable. The smallpox example demonstrates that, even in hindsight, it is hard to know what should be done about DURC research. As a practical matter, it also demonstrates that it is difficult to reach consensus about what to do about DURC research going forward. If a poll were conducted now as to whether smallpox virus should have been sequenced, it is likely that decision would split the opinions of scientists, ethicists, and policymakers, just as every other example of DURC research has.

With all of the uncertainty about DURC, it is not surprising that there is considerable disagreement among security experts about what to be concerned about and why. This was demonstrated empirically in a 2015 Delphi study of US biosecurity experts.[71] Delphi studies aim to draw out and analyze the collective judgments of experts and avoid "groupthink." Given that historical examples of biological weapons use are infrequent and intelligence about bioweapons is hard to come by, judgments about the future threat of biological weapons attacks rest largely on expert opinion. For the study, a group of 59 influential US biosecurity experts were asked to estimate the percentage likelihood of a large-scale biological weapons attack occurring within the next 10 years, in any country—a large-scale attack being defined conservatively as an attack resulting in 100 ill people. Participants' answers about the likelihood of an attack stretched from one end of the scale to the other: from 1 to 100% likelihood, with a mean of 57.5%. Study participants were also asked about whether the perpetrator of a biological weapons attack was more likely to be a nation state actor or terrorist, and while there was much disagreement, an overt attack by a state actor was rated significantly less likely than a terrorist actor. Interestingly, participants used the same information to support divergent views: Pathogen access and technical complexity were both used to support opposite conclusions about whether a state or a nonstate actor is more likely to misuse biological science research to develop and deploy a biological weapon. Overall, a synthetic pathogen was thought to be much less likely to be used as a weapon than an unmodified pathogen in the next 10 years. The diversity of views among experts about the risks of misuse of legitimate scientific information makes it difficult to develop a regulatory system for legitimate dual-use research or to be able to determine the risks and benefits of research that might be a security concern.

Controlling Information—Who Decides Who Can Be Trusted?

In addition to making recommendations about educating scientists on dual-use concerns, the NSABB has become an arbiter for problematic dual-use scientific papers and has offered opinions about whether those papers should be published. They performed this role for the publication of the sequence of the 1918 influenza virus and eventually recommended publication.[72] In 2011, they looked at 2 so-called "gain-of-function" (GOF) influenza research papers, in which pathogens were scientifically altered to introduce traits and functions not found in the world.[67,68] Two groups of researchers in the US and the Netherlands deliberately sought to make H5N1 avian influenza more transmissible between mammals—the function gained in this case was transmissibility—to determine how and whether this transition was likely to occur in nature and which genetic mutations would indicate that the virus was evolving to become more dangerous to humans. Although the original experiments were performed by world-class experts in world-class facilities, synthetic biology techniques could allow such influenza strains to be replicated in labs with less-robust safety systems, health monitoring, and experience. In the end, the White House crafted a path of "doing diligence" on GOF research, involving the NSABB, the National Academy of Sciences, and a research effort into the risks and benefits of the research.[73]

The NSABB looked for a mechanism so that the researchers could share the information that was relevant to public health and influenza disease surveillance and use it appropriately, without lowering the barriers to the work being repeated (for either safety or security reasons). They worried that the publication of the genetic sequence of the modified influenza virus would make it too easy for someone to remake it in the laboratory and release it as a weapon. The NSABB did not want the information classified by a government or governments, as classification would bar many public health workers and experts from making legitimate use of the information. Classification regimes are based on nationality, not expertise or good intent; there are many people who would benefit from the information and put it to beneficent use, including the development of medical countermeasures, but who would not be able to receive a US security clearance. The NSABB was unable to find such a mechanism. The NSABB may have been concerned that H5N1 could be misused and posed a biosecurity risk, but there is no guarantee that either the individuals or governments with legitimate access to the information would use it wisely either.

Who should or can decide who has access? A recent case regarding botulism, and a researcher's assertion that they had discovered a new strain, is a good example of the intricacies of the dual-use information problem. Botulism is an often deadly disease caused by the bacteria *Clostridium botulinum,* and it causes paralysis by blocking neurotransmitters in the body. It is a serious problem for infants and is the reason parents are advised to not feed infants honey, which may contain botulinum spores.[74] There are multiple types of botulism strains that produce an array of toxins, labeled "A" through "G." The toxins are different enough so that antisera against A may not be effective against B, C, or the others. In this case, scientists characterized a new botulinum strain to which available antisera may not be effective and were concerned that the toxin information could be misused. They published the discovery about botulinum H in a scientific journal without the genetic sequence, which would typically be required, citing concerns that the information could be misused.[75,76]

This decision to publish the research without the genetic sequencing data was heralded as an example of the "culture of responsibility" that was wished for in the sciences, to deal with the risks of dual-use information.[77] It was then revealed that neither state nor federal officials had asked for this step to be taken to limit the information that was published.[78] In fact, several government health leaders wanted the botulism information in the public domain, in order to promote more active public health surveillance and stimulate research surrounding the newly discovered strain. The researchers who discovered the new strain then decided not to provide a sample to US government health officials for testing, stating that they would not release it until an antiserum was created for it. This raised questions about whether a new form of botulism had actually been discovered. Without sharing the genetic sequence encoding the toxin, or sharing the sample of the botulinum strain, the claim to have discovered a new botulism strain could not be validated.

In the end, government representatives were able to obtain the strain, and another laboratory published results that existing antisera were able to block the "new" botulism, which turned out to be a hybrid of existent strains.[79,80] Should an individual (or government) have the ability to decide whether this sequence information is better left unknown? Given the advances and increasing accessibility of synthetic biology and other biological techniques and the ability to make results widely accessible, the answer to that question seems now to depend only on whether, and how many, people are aware of and care about the scientific issue at hand.

Synthetic Biology and DURC

Whenever the next DURC event occurs, whether it is in the synthetic biology field or not, it is likely that the debate over what should be done about a new area of research or a new publication will begin anew, and it will be tied to the specifics of the problematic case. Proponents and opponents of the research will place different weight on the variables of the case: the specifics of the research in question, the researchers involved, the urgency of the threat that the research is trying to address, and assessment of the danger that the information could be applied to a biological weapon. These qualities are hard to predict, particularly in global, diversified fields like biological research and biotechnology.

Regardless of whether there is anything tangible that can be done to mitigate the risks in dual-use research, scientists need to know their work may lead to dual-use dilemmas and could spark a debate over whether their work is beneficial to society. Gerald L. Epstein, who has written extensively about the governance of dual-use research, argues, "When scientists are asked whether they have addressed the potential that their research might be misused to inflict deliberate harm, the answer must never be, 'We can't afford to constrain science' or 'You know, we never thought about that.'"[81(p34)] If scientists are prepared for the dual-use issue to arise, they may be able to mitigate the risk of misuse and prevent misunderstandings over why the research was undertaken in the first place. The efforts that have been made to date to increase awareness of dual-use issues need to be expanded, so that even if the next dual-use event does not center on a scientific field that has gone "on alert," like the influenza community, the relevant scientific field has some awareness of dual-use issues. It could be that the next dual-use dilemma arises in synthetic biology, or HIV, or some subset of the biosciences that has not experienced this issue as part of its work.

Journal editors can help raise awareness of dual-use research and biosafety by requiring published papers to include detailed explanations of why the work was undertaken despite the perceived risks, as well as detailed safety information. In 2003, the editors of *Science, Nature,* the *Journal of Virology,* and more than a dozen other scientific publications released a Statement on Scientific Publication and Security, which recommended the development of dual-use review for scientific work and stated that "there is information that, although we cannot now capture it with lists or definitions, presents enough risk of use by terrorists that it should not be published."[82(p1149)] Given the ability to publish more information online than might be available in print, there is no longer any excuse to withhold a full list of all the scientific and safety steps that were taken, thereby encouraging those methods to be adopted. It is currently standard practice that research involving animals and human subjects includes a statement that the study was reviewed by the appropriate institutional committee; it would help if research were to be described as being evaluated by the research institution's biosafety committee and the safety precautions taken were to be described, such as use of vaccines, respirators, or personal protective equipment.

Journal editors have repeatedly stated that they do not want to be in the position of arbitrating whether there are security risks in publication of a scientifically worthy research paper. In fact, this has led to calls that dual-use research of concern be identified at the funding level—that is, much earlier in the process than at the final, publication stage. However, journals can promote awareness of dual-use research concerns and enhance others' understanding of why the research was done in spite of possible risks. And because biological research is inherently surprising, and serendipitous discoveries are part of the process, the role for some dual-use evaluation at the publication stage will inevitably be part of the process.

In addition to dual-use concerns, safety should be a top priority for all scientists, and standards should be enacted for exceptionally consequential research. No matter what is created in the laboratory, and no matter where public and expert opinion falls on the value of the work, the result will be less dangerous if it is properly contained.

Synthetic Biology and Security Conclusions

The risks of misuse are not a complete picture of the role that synthetic biology can play in security. The advances made in synthetic biology could be misused, but they also could be exploited for security gains. For example, the Defense Advanced Research Projects Agency (DARPA) of the DoD is a significant funder of synthetic biology research and applications, spending more than 3 times the amount that is spent by the US National Science Foundation.[83] The projects that are funded are foundational to developing tools in the field but are also for specific applications useful to the DoD, including keeping barnacles from growing on US naval ships, power generation through living cells, extraplanetary resource generation, or speeding up the development of medical countermeasures against emerging infectious diseases here on earth. The speed afforded by new synthetic biology techniques holds promise to bring new medical countermeasures onto the market faster. Distributed manufacturing, also made more achievable through synthetic biology, can then bring those medical countermeasures to more people in need. Safety and security systems, including decoys, can be designed so that they are intrinsic to synthetic organisms; the synthetic organisms themselves can thus foil attempts to either steal them or manipulate their genetic information.[84,85]

Synthetic biology isn't needed to make a biological weapon; there are many plausible scenarios that require less intricate and fallible plans to make a biological weapon. However, synthetic biology does indeed lower barriers to biological weapons development because of the ability to make stocks of a pathogen without starting from an existing strain, and because the technologies developed through synthetic biology increase the accessibility of biological research and manipulations that can be used for harm. In other words, synthetic biology increases our already high vulnerability to biological terrorism and biological warfare, and it joins other areas of science that are dual use and that lower barriers to weaponization. While some policy options are available to mitigate the risks particular to synthetic biology—outreach to the synthetic biology community to raise awareness of the risks, screening gene synthesis orders for harmful sequences—the risks can't be entirely closed off without also sacrificing the benefits that can be gained from the technology, including benefits to security.

The difficulty in entirely preventing a biological weapons attack, whether made using synthetic biology or older techniques, is well documented: a determined adversary has a wide variety of paths to exploit biological materials and information to develop and use a weapon, and while that path can be made more secure, misuse cannot be categorically prevented.[86] However, biological weapons are different from nuclear weapons in that there are more opportunities to prevent mass lethality after an attack. The ability to detect that an event has occurred, respond quickly to limit the spread of disease, and to make countermeasures so that people can be protected from harm is the most reliable path toward limiting the size of a biological weapons event. A suite of technologies like synthetic biology can help to speed the development of these countermeasures, as it should be used to spur advances in the faster detection and treatment of new diseases, whether they come from nature or from man.

References

1. World Health Organization. Ebola Situation Report – 16 March 2016. http://apps.who.int/ebola/current-situation/ebola-situation-report-16-march-2016. Accessed April 1, 2016.
2. The World Bank. World Bank group Ebola response fact sheet. Updated April 6, 2016. http://www.worldbank.org/en/topic/health/brief/world-bank-group-ebola-fact-sheet. Accessed August 24, 2016.
3. Boddie C, Sell TK, Watson M. Federal funding for health security in FY2016. *Health Secur* 2015;13(3):186-206.
4. Hunt M. Dr. Fauci: "nature is the worst terrorist." *cnsnews.com.* October 6, 2014. http://www.cnsnews.com/news/article/melanie-hunter/dr-fauci-nature-worst-terrorist. Accessed August 24, 2016.
5. The world's deadliest bioterrorist. *The Economist* April 28, 2012. http://www.economist.com/node/21553448. Accessed August 24, 2016.
6. Eskenazi J. 'Worst terrorist is nature itself,' says Israeli-American virologist. *JWeekly.com* November 2, 2007. http://www.jweekly.com/article/full/33711/-worst-terrorist-is-nature-itself-says-israeli-american-virologist/. Accessed August 24, 2016.
7. United Nations Office for Disarmament Affairs. Convention on the Prohibition of the Development, Production and Stockpiling of Bacteriological (Biological) and Toxin Weapons and on their Destruction. http://disarmament.un.org/treaties/t/bwc. Accessed August 24, 2016.
8. Torok TJ, Tauxe RV, Wise RP, et al. A large community outbreak of salmonellosis caused by intentional contamination of restaurant salad bars. *JAMA* 1997;278(5):389-395.
9. Danzig R, Sageman M, Leighton T, et al. *Aum Shinrikyo:*

Insights into How Terrorists Develop Biological and Chemical Weapons. Washington, DC: Center for a New American Security; December 2012. http://www.cnas.org/files/documents/publications/CNAS_AumShinrikyo_SecondEdition_English.pdf. Accessed August 24, 2016.
10. Shea DA, Gottron F. *Ricin: Technical Background and Potential Role in Terrorism.* Washington, DC: Congressional Research Service; April 17 2013. https://www.fas.org/sgp/crs/terror/RS21383.pdf. Accessed August 24, 2016.
11. Incidents involving ricin. Wikipedia. 2016. https://en.wikipedia.org/wiki/Incidents_involving_ricin. Accessed April 1, 2016.
12. Herriman R. ISIS and bioterrorism: tularemia planned use in Turkey's water. *Outbreak News Today* January 21, 2016. http://outbreaknewstoday.com/isis-and-bioterrorism-tularemia-planned-use-in-turkeys-water-67823/. Accessed August 24, 2016.
13. Doornbos H, Moussa J. Found: the Islamic state's terror laptop of doom. *Foreign Policy* August 28, 2014. http://foreignpolicy.com/2014/08/28/found-the-islamic-states-terror-laptop-of-doom/. Accessed August 24, 2016.
14. Mowatt-Larssen R. *Al Qaeda Weapons of Mass Destruction Threat: Hype or Reality?* Cambridge, MA: Belfer Center; January 2010. http://bit.ly/2dqQL9a. Accessed August 24, 2016.
15. Synthetic Biology Project. Synthetic biology products and applications inventory. Woodrow Wilson International Center for Scholars; 2016. http://www.synbioproject.org/cpi/. Accessed August 24, 2016.
16. Regalado A. Top U.S. intelligence official calls gene editing a WMD threat. *MIT Technology Review* February 9, 2016. https://www.technologyreview.com/s/600774/top-us-intelligence-official-calls-gene-editing-a-wmd-threat/. Accessed August 24, 2016.

17. Statement for the Record: Worldwide Threat Assessment of the US Intelligence Community. Senate Armed Services Committee. James R. Clapper, Director of National Intelligence. February 9, 2016. http://www.dni.gov/files/documents/SASC_Unclassified_2016_ATA_SFR_FINAL.pdf. Accessed August 24, 2016.
18. Rambhia KJ, Ribner AS, Gronvall GK. Everywhere you look: select agent pathogens. *Biosecur Bioterror* 2011;9(1):69-71.
19. Broad WJ. Geographic gaffe misguides anthrax inquiry. *New York Times* January 30, 2002. http://www.nytimes.com/2002/01/30/national/30AMES.html. Accessed August 24, 2016.
20. Oldstone MBA. *Viruses, Plagues, and History: Past, Present, and Future*. New York: Oxford University Press; 2010.
21. Massung RF, Liu LI, Qi J, et al. Analysis of the complete genome of smallpox variola major virus strain Bangladesh-1975. *Virology* 1994;201(2):215-240.
22. Nalca A, Rimoin AW, Bavari S, Whitehouse CA. Reemergence of monkeypox: prevalence, diagnostics, and countermeasures. *Clin Infect Dis* 2005;41(12):1765-1771.
23. Esposito JJ, Knight JC. Orthopoxvirus DNA: a comparison of restriction profiles and maps. *Virology* 1985;143(1):230-251.
24. Shchelkunov SN, Totmenin AV, Babkin IV, et al. Human monkeypox and smallpox viruses: genomic comparison. *FEBS Lett* 2001;509(1):66-70.
25. Zeliadt N. Pox swap: 30 years after the end of smallpox, monkeypox cases are on the rise. *Scientific American* August 31, 2010. http://www.scientificamerican.com/article/pox-swap-30-years-after-small-pox-monkey-pox-on-the-rise/. Accessed August 24, 2016.
26. World Health Organization. *Scientific Review of Variola Virus Research 1999-2010.* December 2010.

http://whqlibdoc.who.int/hq/2010/WHO_HSE_GAR_BDP_2010.3_eng.pdf. Accessed August 24, 2016.

27. Gronvall GK, Ravi S, Cicero A, Inglesby T. *Singapore-US Strategic Dialogue on Biosecurity.* UPMC Center for Health Security. December 2014. http://www.upmchealthsecurity.org/our-work/pubs_archive/pubs-pdfs/2014/Singapore%20Report.pdf. Accessed August 24, 2016.
28. Maurer SM, Lucas KV, Terrell S. *From Understanding to Action: Community-Based Options for Improving Safety and Security in Synthetic Biology.* University of California, Berkeley, Richard & Rhoda Goldman School of Public Policy. April 15, 2006. http://citeseerx.ist.psu.edu/viewdoc/download?doi=10.1.1.132.8678&rep=rep1&type=pdf. Accessed August 24, 2016.
29. Service RF. The synthetic biologist's code. *Science* May 23, 2006. http://news.sciencemag.org/2006/05/synthetic-biologists-code. Accessed August 24, 2016.
30. Synthetic Biology Project. Woodrow Wilson International Center for Scholars. 2016. http://www.synbioproject.org/about/. Accessed April 10, 2016.
31. You E. FBI connects science and security communities. *Science, Safety, and Security Quarterly* May 2012;(2):1-5. http://www.phe.gov/s3/Documents/s3newsletter-may2012.pdf. Accessed August 24, 2016.
32. Lempinen EW. FBI, AAAS collaborate on ambitious outreach to biotech researchers and DIY biologists. April 1, 2011. http://www.aaas.org/news/fbi-aaas-collaborate-ambitious-outreach-biotech-researchers-and-diy-biologists. Accessed August 24, 2016.
33. Cello J, Paul AV, Wimmer E. Chemical synthesis of poliovirus cDNA: generation of infectious virus in the absence of natural template. *Science* 2002;297(5583):1016-1018.

34. Tumpey TM, Basler CF, Aguilar PV, et al. Characterization of the reconstructed 1918 Spanish influenza pandemic virus. *Science* 2005;310(5745):77-80.
35. Gibson DG, Glass JI, Lartique C, et al. Creation of a bacterial cell controlled by a chemically synthesized genome. *Science* 2010;329(5987):52-56.
36. Hutchison CA 3rd, Chuang RY, Noskov VN, et al. Design and synthesis of a minimal bacterial genome. *Science* 2016;351(6280):aad6253.
37. Randerson J. Revealed: the lax laws that could allow assembly of deadly virus DNA. *Guardian* June 14, 2006. http://www.theguardian.com/world/2006/jun/14/terrorism.topstories3. Accessed August 24, 2016.
38. Correspondence from May 2006 between James Randerson and Drew Endy regarding the possibility of demonstrating that it is possible to mail-order synthesize the DNA encoding human pathogens. 2006. http://bit.ly/2dMu7Ll. Accessed August 24, 2016.
39. Carlson R. Comments on mail-ordering smallpox genes. Vol 20132006. http://www.synthesis.cc/2006/06/comments-on-mail-ordering-smallpox-genes.html#comments
40. Jones R. Sequence screening. In: Garfinkel MS, Endy D, Epstein GL, Friedman RM, eds. *Working Papers for Synthetic Genomics: Risks and Benefits for Science and Society.* J. Craig Venter Institute, CSIS, MIT; 2007. http://www.jcvi.org/cms/fileadmin/site/research/projects/synthetic-genomics-report/Commissioned-Papers-Synthetic-Genomics-Governance.pdf. Accessed August 24, 2016.
41. International Gene Synthesis Consortium (IGSC). Harmonized Screening Protocol: Gene Sequence & Customer Screening to Promote Biosecurity. 2009. http://www.genesynthesisconsortium.org/wp-content/uploads/2012/02/IGSC-Harmonized-Screening-Protocol1.pdf.

42. Randerson J. Did anyone order smallpox? *Guardian* June 23, 2006. http://www.theguardian.com/science/2006/jun/23/weaponstechnology.guardianweekly. Accessed August 24, 2016.
43. Department of Health and Human Services. Screening framework guidance for providers of synthetic double-stranded DNA. *Federal Register* October 13, 2010;75(197):62820-62832. http://www.gpo.gov/fdsys/pkg/FR-2010-10-13/html/2010-25728.htm. Accessed August 24, 2016.
44. Wadman M. US drafts guidelines to screen genes. *Nature* December 4, 2009. http://www.nature.com/news/2009/091204/full/news.2009.1117.html. Accessed August 25, 2016.
45. Gronvall GK. HHS guidance on synthetic DNA is the right step. *Biosecur Bioterror* 2010;8(4):373-376.
46. Gronvall GK. *Mitigating the Risks of Synthetic Biology.* Council on Foreign Relations. February 2015. http://www.cfr.org/health/mitigating-risks-synthetic-biology/p36097. Accessed August 25, 2016.
47. Carter SR, Friedman RM. *DNA Synthesis and Biosecurity: Lessons Learned and Options for the Future.* J. Craig Venter Institute. October 2015. http://www.jcvi.org/cms/fileadmin/site/research/projects/dna-synthesis-biosecurity-report/report-complete.pdf. Accessed August 25, 2016.
48. Office of the Director of National Intelligence, Intelligence Advanced Research Projects Activity (IARPA). Functional Genomic and Computational Assessment of Threats (Fun GCAT). 2016. https://www.iarpa.gov/index.php/research-programs/fun-gcat. Accessed July 13, 2016.
49. Danzig R. *Catastrophic Bioterrorism—What Is To Be Done?* Washington, DC: Center for Technology and National Security Policy, National Defense University. August 2003.

http://www.response-analytics.org/images/Danzig_Bioterror_Paper.pdf. Accessed August 25, 2016.

50. Gilchrist CA, Turner SD, Riley MF, Petri WA, Jr, Hewlett EL. Whole-genome sequencing in outbreak analysis. *Clin Microbiol Rev* 2015;28(3):541-563.
51. Rasko DA, Worsham PL, Abshire TG, et al. Bacillus anthracis comparative genome analysis in support of the Amerithrax investigation. *Proc Natl Acad Sci U S A* 2011;108(12):5027-5032.
52. Hessel A, Goodman M, Kotler S. Hacking the president's DNA. *Atlantic* November 2012. http://www.theatlantic.com/magazine/archive/2012/11/hacking-the-presidents-dna/309147/. Accessed August 25, 2016.
53. Church G, Casci J, Brin D, Maria CS. Sci-Fi Political Thriller, Or Fact? In: Soboroff J, ed. *HuffPost Live*2012. https://www.amazon.com/Hacking-Obamas-DNA-Political-Thriller/dp/B00TNHS7HO
54. Bowen CD, Renner DW, Shreve JT, et al. Viral forensic genomics reveals the relatedness of classic herpes simplex virus strains KOS, KOS63, and KOS79. *Virology* 2016;492:179-186.
55. Centers for Disease Control and Prevention. Coughing and sneezing. Updated December 28, 2009. http://www.cdc.gov/healthywater/hygiene/etiquette/coughing_sneezing.html. Accessed August 25, 2016.
56. Starobin P. The accidental autocrat. *Atlantic* March 2005. http://www.theatlantic.com/magazine/archive/2005/03/the-accidental-autocrat/303725/. Accessed August 25, 2016.
57. United Nations Security Council Resolution 1540. http://www.un.org/en/sc/1540/. Accessed August 25, 2016.
58. Centers for Disease Control and Prevention. Federal Select Agent Program. Select agents and toxins list. 2014.

http://www.selectagents.gov/SelectAgentsandToxinsList.html. Accessed August 25, 2016.
59. Uniting and Strengthening America by Providing Appropriate Tools Required to Intercept and Obstruct Terrorism (USA PATRIOT ACT) Act of 2001, § 817. Expansion of the Biological Weapons Statute (2001).
60. Jackson RJ, Ramsay AJ, Christensen CD, Beaton S, Hall DF, Ramshaw IA. Expression of mouse interleukin-4 by a recombinant ectromelia virus suppresses cytolytic lymphocyte responses and overcomes genetic resistance to mousepox. *J Virol* 2001;75(3):1205-1210.
61. Committee on Research Standards and Practices to Prevent the Destructive Application of Biotechnology;Development, Security, and Cooperation; Policy and Global Affairs; National Research Council. *Biotechnology Research in an Age of Terrorism*. Washington, DC: National Academies Press; 2004.
62. US Department of Health and Human Services. National Institutes of Health. United States Government Policy for Oversight of Life Sciences Dual Use Research of Concern. March 29, 2012. http://osp.od.nih.gov/office-biotechnology-activities/dual-use-reasearch-concern-policy-information-national-science-advisory-board-biosecurity-nsabb/united-states-government-policy-oversight-life-sciences-dual-use-research-concern. Accessed August 25, 2016.
63. Vogel KM. Expert knowledge in intelligence assessments: bird flu and bioterrorism. *International Security* 2014;38(3):39-71. http://belfercenter.hks.harvard.edu/files/IS3803_pp039-071.pdf. Accessed August 25, 2016.
64. Athamna A, Athamna M, Abu-Rashed N, Medlej B, Bast DJ, Rubinstein E. Selection of Bacillus anthracis isolates resistant to antibiotics. *J Antimicrob Chemother* 2004;54(2):424-428.
65. Brook I, Elliott TB, Pryor HI 2nd, et al. In vitro resistance

of Bacillus anthracis Sterne to doxycycline, macrolides and quinolones. *Int J Antimicrob Agents* 2001;18(6):559-562.
66. Pomerantsev AP, Staritsin NA, Mockov Yu V, Marinin LI. Expression of cereolysine AB genes in Bacillus anthracis vaccine strain ensures protection against experimental hemolytic anthrax infection. *Vaccine* 1997;15(17-18):1846-1850.
67. Russell CA, Fonville JM, Brown AE, et al. The potential for respiratory droplet-transmissible A/H5N1 influenza virus to evolve in a mammalian host. *Science* 2012;336(6088):1541-1547.
68. Imai M, Watanabe T, Hatta M, et al. Experimental adaptation of an influenza H5 HA confers respiratory droplet transmission to a reassortant H5 HA/H1N1 virus in ferrets. *Nature* 2012;486(7403):420-428.
69. Dando M. Advances in neuroscience and the Biological and Toxin Weapons Convention. *Biotechnol Res Int* 2011;2011:973851.
70. Soll SJ, Neil SJ, Bieniasz PD. Identification of a receptor for an extinct virus. *Proc Natl Acad Sci U S A.* 2010;107(45):19496-19501.
71. Boddie C, Watson M, Ackerman G, Gronvall GK. Biosecurity. Assessing the bioweapons threat. *Science* 2015;349(6250):792-793.
72. Kennedy D. Better never than late. *Science* 2005;310(5746):195.
73. White House Office of Science and Technology Policy. Doing diligence to assess the risks and benefits of life sciences gain-of-function research. October 17, 2014. http://www.whitehouse.gov/blog/2014/10/17/doing-diligence-assess-risks-and-benefits-life-sciences-gain-function-research. Accessed August 25, 2016.
74. Rosano C, Bennett DA, Newman AB, et al. Patterns of focal gray matter atrophy are associated with bradykinesia and

gait disturbances in older adults. *J Gerontol A Biol Sci Med Sci* 2012;67(9):957-962.
75. Dover N, Barash JR, Hill KK, Xie G, Arnon SS. Molecular characterization of a novel botulinum neurotoxin type H gene. *J Infect Dis* 2014;209(2):192-202.
76. Barash JR, Arnon SS. A novel strain of Clostridium botulinum that produces type B and type H botulinum toxins. *J Infect Dis* 2014;209(2):183-191.
77. Casadevall A, Enquist L, Imperiale MJ, Keim P, Osterholm MT, Relman DA. Redaction of sensitive data in the publication of dual use research of concern. *MBio* 2014;5(1):e00991-00913.
78. Greenfieldboyce N. Who's protecting whom from deadly toxin? *National Public Radio* April 21, 2014. http://www.npr.org/blogs/health/2014/04/21/305650796/whos-protecting-whom-from-deadly-toxin. Accessed August 25, 2016.
79. Fan Y, Barash JR, Lou J, Conrad F, Marks JD, Arnon SS. Immunological characterization and neutralizing ability of monoclonal antibodies directed against botulinum neurotoxin type H. *J Infect Dis* 2016;213(1):1606-1614.
80. Maslanka SE, Lúquez C, Dykes JK, et al. A novel botulinum yoxin, previously reported as serotype H, has a hybrid structure of known serotypes A and F that is neutralized with serotype A antitoxin. *J Infect Dis* 2016;213(3):379-385.
81. Epstein GL. Preventing biological weapon development through the governance of life science research. *Biosecur Bioterror* 2012;10(1):17-37.
82. Atlas R, Campbell P, Cozzarelli NR, et al. Statement on scientific publication and security. *Science* 2003;299(5610):1149.
83. Office of Technical Intelligence, Office of the Assistant Secretary of Defense for Research and Engineering. *Technical Assessment: Synthetic Biology.* Department of Defense Research and Engineering. January 2015.

http://defenseinnovationmarketplace.mil/resources/OTI-SyntheticBiologyTechnicalAssessment.pdf. Accessed August 25, 2016.
84. Cai Y, Agmon N, Choi WJ, et al. Intrinsic biocontainment: multiplex genome safeguards combine transcriptional and recombinational control of essential yeast genes. *Proc Natl Acad Sci U S A* 2015;112(6):1803-1808.
85. Krishnakumar R. Intrinsic biocontainment: state of the science and future possibilities. Paper presented at: SB 6.0: The Sixth International Meeting on Synthetic Biology; 2013; Imperial College, London, England.
86. Graham B, Talent J. Bioterrorism: redefining prevention. *Biosecur Bioterror* 2009;7(2):125-126.

CHAPTER 3.

ON SAFETY

While the benefits of synthetic biology are likely to be significant and far-ranging, the increased power and widespread accessibility of these biological technologies also raise concerns about their safety. To date, synthetic biology has *not* been associated with any accidents either inside or outside of a laboratory. However, the possibility for negative consequences as a result of a biosafety lapse with these powerful tools needs to be examined, planned for, and mitigated. This chapter describes the particular safety risks that these synthetic biology technologies may pose to people, animals, and the environment and proposes policies intended to prevent and mitigate those risks.

There are several main categories for safety concerns about synthetic biology that will be discussed in this chapter. The first category involves "outside the laboratory" synthetic biology applications. For these uses of synthetic biology, synthetic organisms are designed to be deliberately released into the environment. Applications that fall into this category include mosquito control, agriculture, pollution remediation, mining, biofuels, medications, or even the re-creation of extinct animals.[1] These endeavors could yield unintended and accidental consequences if biosafety risks are not addressed and carefully thought through.

The second concern for safety relates to the experience level of synthetic biology practitioners with biosafety and basic principles of containment. Practitioners of synthetic biology may not have years of experience and training in laboratories where biosafety concerns are immediate, such as in infectious disease laboratories, and may not know how to take the correct precautions to protect themselves and others.

The final category of biosafety concern for synthetic biology is a general concern for advanced bioscience: Because of the increased access to powerful technologies to more laboratories and amateurs in the world, consequential bio-errors may occur more frequently. If an accident occurs with a transmissible pathogen, the damaging effects of an accident could spread well beyond the laboratory.

While some technical, policy, and regulatory steps have been taken to address each of these concerns, much more needs to be done to mitigate biosafety risks in synthetic biology. Doing so will not only minimize harm to people, animals, and the environment but also allow the long-term promise of synthetic biology to be realized.

Outside of the Laboratory Concerns

Many synthetic biology applications keep synthetic organisms in containment. For example, synthetic organisms may be kept in closed manufacturing systems, or contained within a laboratory equipped to prevent organisms from escaping into the outside environment. However, some synthetic biology applications actually require organisms that have been modified by synthetic biology to be placed outside of the traditional laboratory environment and outside traditional containment. For example, synthetic bacteria designed for environmental cleanup or remediation would need to be placed in the area they would be cleaning; mining, pollution detection, and agricultural uses of synthetic organisms pose similar challenges. Even some medical applications that are being developed pose new "outside the laboratory" biosafety concerns, such as using engineered microbes to treat Crohn's disease or oral inflammation, or to modify pigs' organs so that they can be used for transplantation into humans without being rejected.[2-4]

With these synthetic organisms outside of traditional containment, it raises the possibility that they could interact with "natural" organisms or affect the environment in unintended ways. New methods are needed to contain the synthetic organisms outside of traditional laboratory containment so that the modified organisms do the job they were designed for and nothing else. There are precedents for genetically modified organisms (GMOs) to be used in agriculture and food, and there are regulatory systems and oversight tailored to that purpose, but some of these new applications that are possible due to synthetic biology do not readily fall into the same oversight mechanisms designed for agriculture or medicine, and so the standards and regulatory environment are less clear.

Biosafety by Design

One approach to increase containment of synthetic organisms outside the laboratory is called "intrinsic" biosafety—that is, the biosafety is built into the organism, so that the synthetic organisms can't escape boundaries that are set for them. Some forms of intrinsic biosafety involve engineering organisms so that they are not able to survive without specific human intervention, such as by supplying a nutrient that is essential for life. If the nutrient is not supplied, the synthetic organism will die, and it will stay "contained" in the area where that essential nutrient is supplied.

Early attempts to develop intrinsic biosafety often had a single point of failure—and often failed, leading to the organism's escape. The synthetic organisms that required an essential ingredient to survive simply evolved so that the supposedly essential nutrient was no longer necessary. Instead of relying on a single point of failure, newer intrinsic biosafety approaches that are crafted with synthetic biology are more subtle and rely on combinatorial complexity to boost the number of points that would need to fail before the organism could escape containment. This drastically reduces the rates of failure and escape. Furthermore, the synthetic organisms can be nearly as robust as the wild-type (natural) organisms. In the past, engineered organisms were often slow to grow and thus undesirable to use in agriculture or manufacturing.[5]

Future efforts to increase intrinsic biosafety may involve entirely different life forms—what are called "orthogonal life forms"—which use artificial genetic languages (no A, C, G, and T) and require specific nutrients that are not found in nature at all.[6] An orthogonal approach decreases the risks that the synthetic organisms would be able to evolve to survive without deliberate human intervention, that wild-type organisms would be able to use the nutrient and would prevent the transfer of synthetic traits to natural biological systems.[7,8]

An intrinsic biosafety approach was demonstrated in 2013, using the traditional laboratory workhorse *Escherichia coli* (*E. coli*) bacteria. Scientists recoded the bacterial DNA so that they were able to incorporate a nonstandard amino acid into its proteins. Basically, scientists expanded the kinds of proteins the *E. coli* could make.[9] The researchers then took what they called their "genomically recoded organism," or GRO, and increased its biosafety. They modified it to become dependent on a synthetic amino acid that the GRO couldn't make itself. They added 49 mutations that make the organism dependent on the artificial nutrient, which lowers the risk of "escape" mutations; they grew

1 trillion of the *E. coli* cells, and none escaped.[10,11] George Church, a professor at Harvard and one of the senior researchers in this work, explained that they were simply applying common engineering principles to this biological problem: "If you make a chemical that's potentially explosive, you put stabilizers in it. If you build a car, you put in seat belts and airbags."[11] Engineering synthetic biology products can include highly engineered safety mechanisms.

In another demonstration of intrinsic biosafety, scientists incorporated a containment system into synthetic *Saccharomyces cerevisiae,* or Brewer's yeast; this is another laboratory workhorse as well as an industrially important organism.[5] Scientists developed a synthetic yeast strain that contains 2 unrelated safeguard switches, which can be paired with special essential nutrients, so that a designer microbe could be kept alive by "supplying a blend of such compounds as a kind of 'special sauce.'"[5(p1803)] The escape rates for the individual safeguards were found to be 10^{-8} to 10^{-6}, which leads to a vanishingly small risk of escape with 2 safeguards built into the organism.[5]

Intrinsic biosafety work has other benefits besides safety, including the protection of intellectual property. The ability to put a signature or watermark into the DNA of a synthetic organism could identify the designers of the organism. Such a watermark was put into *Mycoplasma mycoides* JCVI-syn1.0, "the first self-replicating species that we have had on the planet whose parent is a computer," designed by researchers from the J. Craig Center Institute.[12(pA17)] At the announcement that he and his team had made the first synthetic bacterial cell, Venter also revealed that the cell had coded messages placed in the genome. If decoded, one could read the names of all 46 authors on the publication, other key contributors, an email address, and some quotations, including one from the physicist Richard Feynman: "What I cannot create, I cannot understand."[13,14] In the future, decoy circuits can be put into organisms that would confuse those who intend to steal synthetic organisms, or to make the organism self-destruct if certain harmful manipulations are attempted.[5]

For safety and security reasons, funding new types of intrinsic biosafety approaches should be a priority—the research won't be pursued if it isn't funded. Right now, this work is largely being pursued by the Department of Defense's Defense Advanced Research Projects Agency (DARPA) but should also be transferred into other US government funding streams, as DARPA does not typically invest in technical solutions beyond a short time period.[15]

In addition to growing the field of intrinsic safety and other new approaches to containment, however, it would be helpful if there were a standard analytical method to determine the safety of a newly designed synthetic organism.[7] To what safety standard should new synthetic organisms adhere, and using which criteria? Analysts from the Woodrow Wilson International Center for Scholars suggest criteria based on how well synthetic organisms interact with natural ones; how well synthetic organisms survive in their intended environment; their capability to evolve; and their ability to adapt to new environmental niches.[16] Determining appropriate safety criteria is an area that should be thought about well before a problem arises.

Currently, the US government is undertaking a process to update the Coordinated Framework, which guides the regulation and oversight of biotechnology products by the FDA, USDA, and the EPA.[17] Updating the framework is an opportunity to provide appropriate oversight for new synthetic biology techniques—and the framework appears to need updating. A 2013 fundraising campaign on Kickstarter caused a great deal of consternation by promising to produce glowing plants and distribute seeds to more than 8,000 supporters.[18] The biological mechanisms they intended to use to produce the plants, distribute them, and plant them do not violate any current rules or regulations; however, allowing glowing plants to be introduced into the environment without regulatory review has struck many scientists as inappropriate.[19] At this time, however, it is only a theoretical concern: The technical details of producing glowing plants have proven to be far more complex than envisioned at the outset of the venture, and no plants have been distributed yet. In fact, the company is pivoting from glowing plants to produce patchouli-scented moss, which they believe will be an easier product to produce.[20] Nonetheless, policy experts at the J. Craig Venter Institute concluded that, while US regulatory agencies have adequate legal authority to address most concerns that synthetic biology may pose, there is a need to update regulations.[21] Glowing plants may (eventually) just be the beginning: De-extinction—a goal for some synthetic biologists, to bring "back" extinct animals, including the wooly mammoth—is likely to raise additional safety, as well as ethical, concerns.[22]

Use of Synthetic Biology to Get Rid of the Mosquito

One outside-the-laboratory application of synthetic biology has been suggested as a solution to the 2016 Zika epidemic: eliminating *Aedes aegypti,* the mosquito blamed for Zika virus transmission.[23] The Zika virus has been around for decades—it was originally isolated in 1947 from the Zika forest in Uganda.[24] But there has been an explosive increase in the number of illnesses, primarily in Brazil. Preliminary estimates are of 440,000 to 1.3 million cases reported worldwide through the end of December 2015.[25] Dengue and chikungunya are also transmitted by *Aedes aegypti* mosquitoes; both of these diseases are associated with higher levels of mortality. But Zika has an association with microcephaly, a syndrome in which babies are born with smaller heads and damaged or incompletely developed brains. In Brazil, 8,451 cases of microcephaly and other congenital malformations of the central nervous system (CNS) were reported over a 10-month period in 2015-16 in newborns.[26] Women were asked to delay pregnancy, and the US CDC advised pregnant women to avoid traveling to Brazil for the 2016 Olympic games.[27,28] A vaccine is not likely to be available for at least 3 years or more.[29]

Use of synthetic biology to make a "gene drive" could theoretically drive *Aedes aegypti* mosquitoes to extinction, at least in that geographic area. Gene drives engineer non-Mendelian inheritance, so that instead of a specific gene being passed along from a parent to *half* of offspring as is typical, almost *all* offspring inherit the gene.[30] The synthetic biology tool drives inheritance. To develop a gene drive, scientists would use a newly developed gene-editing technique, clustered regularly interspaced short palindromic repeats (CRISPR/Cas9). The gene drive would be designed so that offspring of the engineered mosquitoes would be limited in their ability to reproduce; for example, the drive might lead all offspring to be male, and so the species would—theoretically—die out.

Before Zika, some scientists had proposed using a gene drive to make *Anopheles* mosquitoes resistant to malaria.[31] Large numbers of engineered mosquitoes would be released to mate with the malaria-sensitive population, resulting in malaria-resistant offspring, which would pass along malaria resistance.[32] If successful, such a project could decrease the prevalence of a disease that currently kills more than 600,000 people—mostly children—each year. In 2015 alone, malaria killed more than 305,000 African children before their fifth birthday.[33] Yet, while the public health benefits could be enormous, the resistance genes that would be synthetically added to the mosquitoes could theoretically drift to other species, causing unknown and hard-to-reverse consequences to other species and to the environment.[34]

Aedes aegypti is an invasive species in the Americas, and its introduction has clearly caused harm. Eliminating the insect is not *likely* to hurt any other life form, but it is hard to guarantee, and ecological interactions may be difficult to predict.[35] Even with an invasive species, there may be a useful biological niche that is being filled. It is also unknown whether the changes would be stable over many generations.[36] And a gene drive might be difficult to reverse if something went wrong.

There are 3 US laboratories that do research on mosquitoes, that are working toward a gene drive that would eliminate *Aedes aegypti,* and that may have candidates within a year.[35] Deciding whether and how to safely proceed will likely take longer than that, and more time will be needed for safety and efficacy tests. There would be tests in the laboratory, where biocontainment of the modified mosquitoes is heavily controlled; in one insectary at University of California (Irvine), mosquitoes are contained behind 5 sets of doors, and testing uses only a species that doesn't survive in California, should the mosquitoes escape.[36] If the laboratory tests prove successful, the mosquitoes might then be tested in an environmentally secluded area, such as an island.[35]

The development of regulations to limit the use of this technology, as well as public debate about whether the *Aedes aegypti* mosquito should be eliminated, will also take time. There has already been some thinking about this; synthetic biologists have taken the lead in thinking about the safety consequences and have been developing a series of commonly agreed on safeguards for laboratory research into gene drives. For example, there could be a combination of multiple stringent confinement strategies, since any single confinement strategy could fail.[37] Scientists have also put forward potential guidelines for how to safely use them outside of the laboratory.[34] A National Academy of Sciences panel recommended that it is too soon for a release of gene drives into the environment, but that the potential benefits are significant, justifying proceeding with laboratory research to fill knowledge gaps and to eventually conduct highly controlled field trials.[38]

There are other options for mosquito control that can and have worked in the past. Still, these methods are extraordinarily difficult to implement successfully. Personal liberties and property rights can be put at odds with exhaustive mosquito control, and the manpower needed to accomplish it is enormous.[39] *Aedes aegypti* prefer urban areas, bite in the day so bed nets have little effect, and can breed in small containers of water—even in discarded soda caps.[39] Brazil sent 220,000 soldiers door-to-door to check for mosquitoes breeding in old tires and swimming pools, an exhausting effort to maintain. There have also been calls to use DDT, a powerful pesticide banned in many countries because of the ecological damage it can cause, as famously documented in the 1962 book *Silent Spring*.[40]

There are additional biological control measures being used to combat mosquitoes, though these methods have less power than a gene drive. One method is already in place in Brazil to fight dengue, which is also transmitted by *Aedes aegypti:* Oxitec, a British company, has released many mosquitoes that are engineered to pass a lethal gene to offspring. In small tests, this approach has lowered mosquito populations by 80% or more.[39] Another approach is to infect mosquitoes with Wolbachia, a bacterium that does not infect them naturally and that limits the ability of the mosquitoes to pick up and transmit other viruses. The bacteria are passed from a parent mosquito into the next generation, so the bacterial infection is promulgated.[39] This project is supported by the Bill and Melinda Gates Foundation, and tests are under way in Indonesia and Vietnam to see if the technique can reduce the number of people infected with dengue.[39]

New People in New Types of Laboratories

The second category of concern for synthetic biology biosafety risks stems from the training of the people performing it. Typically, laboratory skills including biosafety are developed over years of training and mentorship in biological research laboratories. Yet, with the advent of synthetic biology, some people have been able to pursue biological research outside of a traditional laboratory, and they may not have taken a traditional path to biosciences work, with exposure to biosafety concepts. In response, biosafety training needs to be adapted so that the people who need safety advice have an opportunity to receive it.

One reason practitioners of synthetic biology may not have received biosafety training is because they are performing research in nontraditional laboratories. DIY ("do-it-yourself") Bio falls into this category. DIY Bio is no longer a new phenomenon: In 2005, synthetic biology and biotechnology industry analyst Rob Carlson wrote in *Wired* that "the era of garage biology is upon us" and that a few thousand dollars of investment can allow someone to start "hacking biology."[41] Years later, costs have dropped, accessibility has increased, and the DIY Bio movement has grown: There are more than 4,400 subscribers to the online forum, where DIY Bio practitioners communicate about their work and arrange to meet up with other "bio-hackers" in their communities.[42]

Many of the DIY Bio activities are expressly educational, fun, and hold safety as a high priority. In fact, the organization DIYbio.org was created in 2008 to establish a "vibrant, productive and safe community of DIY biologists."[43] Yet, while most DIY Bio projects are not sophisticated, the tools to do advanced work are accessible: For about $150, an amateur can purchase a gene-editing kit featuring CRISPR—the technology that is revolutionizing the biosciences right now and which is only a few years old.[44]

Community laboratories have endeavored to promote biosafety outreach and bring concepts of containment to the DIY Bio community. There are community laboratories in Brooklyn, San Francisco, Baltimore, and Manchester, England, as well as community groups all over the world, which are as far flung as New Delhi, Tel Aviv, Sao Paulo, and Houston.[43] The rules for community labs tend to be strict: The Baltimore Underground Science Space (BUGSS) requires that those who use the laboratory space abide by certain rules that limit the biosafety risks, such as operating at Biosafety Level 1 (BSL-1), the lowest level of containment.[45] Blood samples or samples taken from natural sources—which could contain pathogens—are not allowed at BUGSS. The safety guidance also acknowledges that "To some the idea of a community laboratory is scary so we must strive to be shining examples of safe and responsible laboratory practice."[45] The safety risks extend to the chemicals that are allowed in the facility, as well as basic, common-sense laboratory safety measures: No eating is allowed in the laboratory.

The DIY Bio community may not have immediate access to traditional academic laboratories, but they do need biosafety advice. To promote biosafety outreach to the DIY Bio community, there is an "ask a biosafety expert" program freely available on the DIY Bio website, currently funded by the Alfred P. Sloan Foundation and supported by the Woodrow Wilson International Center for Scholars.[46] Questions are addressed by biosafety experts, such as "What does a DIY Biologist wear?" (the answer: *It depends on what you are doing*) or "I was growing fecal samples on trypticase soy agar plates, and I wonder whether these could be potentially toxic to me if I'm keeping them in my room?" (the answer: *Yes*).[47]

Unfortunately, funding for this valuable safety resource is drawing to a close. The Alfred P. Sloan Foundation typically provides seed funding for only a few years for areas of need and has now ceased funding projects related to synthetic biology.[48]

Given the importance of this resource and the continual worldwide growth of amateur biology, the US government should allocate resources to continue and expand the amateur outreach programs, to provide biosafety expert advice to those who need it. The biggest expense for the program was for insurance, which was needed for the advice-givers; the Wilson Center insured the program with Lloyds of London. One possibility is that this program could be offered by land grant colleges and universities, as it is part of their mission to provide public access to their expertise.[49]

Another example of a nontraditional laboratory environment is iGEM, the International Genetically Engineered Machine Competition.[50] iGEM began as an in-class competition between Massachusetts Institute of Technology students in 2003, to build synthetic biological systems from standard interchangeable parts, called BioBricks, and operate them in living cells. Biosafety is already well integrated into the iGEM competition; biosafety advice is given to team members, types of projects allowed are strictly limited to keep the risk level low, and a safety committee evaluates all of the iGEM projects. Yet, what is most striking about iGEM is the level of competition: Although iGEM projects are carried out by students, many of them entirely new to bioscience, the projects have been sophisticated. As one example, the 2014 grand prize winner, from Heidelberg, designed a tool box to circularize proteins to make them more physically stable.[51] The work demonstrates the accessibility and the democratization of powerful synthetic biology tools.[51] The competitors may not have had prior biosafety training, but the work is typically performed in academic laboratories, and the students are required to have a mentor, who is usually a scientist who works in a traditional laboratory environment.

The fact that a relatively untrained individual could perform complex bioengineering has triggered concerns from policymakers and the public, and mechanisms to improve the safety and knowledge of the amateur community's activities have already been put into place. An FBI "see something, say something" campaign performs outreach to both the DIY Bio community and to iGEM, though its emphasis is primarily on security versus safety.[52,53] Many commercial DNA synthesis suppliers, often used for synthetic biology projects, screen their orders. This is so that swaths of the genetic material of regulated pathogens—including the causative agents of anthrax, smallpox, or rinderpest—are not able to be purchased without the proper clearances.[54] As commercial synthesis is not the only method for getting DNA, there are limits to how well the screening program can work, but screening does raise barriers to accidental and deliberate misuse.

Even in traditional academic laboratory environments, biosafety outreach and training needs are changing; many synthetic biologists come to the field not through biology but from other technical disciplines, especially engineering and computer science. One noteworthy example is Tom Knight, the founder of the synthetic biology company Ginkgo Bioworks, who was also one of the developers of ARPANET, the precursor of the internet. He has more than 30 patents in computer science and electrical engineering, but as a full professor at MIT, he began to be interested in biology and became a student again, taking molecular biology courses and "learning how to pipette and work in the lab."[55] There are many others who have followed his example, who have gone from electrical engineering, chemical engineering, or computer science to biology; after all, the aims of synthetic biology are to bring engineering principles to the biological world. While Tom Knight took care to learn the safety aspects of his adopted field, others may not be as attentive. At the 2015 American Biological Safety Association meeting, one university biosafety officer urged his colleagues to become more aware of their chemistry and engineering departments, as some of those faculty are likely pursuing biological projects and could use biosafety expertise. Biosafety issues are not limited to the biology and medicine departments anymore.

There are several reasons why this category of biosafety concerns for synthetic biology may not actually be as alarming as it may seem at first: cutting and pasting together strings of DNA has been going on for decades in molecular biology and has thus far been proven exceptionally safe. The vast majority of synthetic biologists are unlikely to be working in areas of research that could cause a problem that would spread beyond a laboratory, such as work in infectious diseases. In addition, if a synthetic biology researcher is part of a research university, receives US government grants, or is planning to submit a product for consideration to a regulatory agency such as the FDA or EPA, he or she will intersect with biosafety and biosecurity regulations. They will likely find themselves in contact with biosafety resources, such as an institutional biosafety officer to provide technical and regulatory support. The synthetic biologist in a university may also have to submit research proposals to an institutional biosafety committee that can provide oversight.

Nonetheless, these shifts in *who* needs biosafety training and *where* they can receive it demonstrate that biosafety training coverage is uneven. This should inspire new methods for outreach. The "Ask a Biosafety Expert" site is a terrific and creative initiative—but there need to be many such plans to bring biosafety to where biological research is being performed, with training and guidance that is appropriate for different types of students. Biosafety training is always a challenge, whether for synthetic biology or any another bioscience area; there is variability in retraining when scientists and laboratory workers move from one institution to another, and like any education program, there is always a new group of people who need training, who are not familiar with established rules and regulations. For example, the World Health Organization has had to notify several laboratories that their published research demonstrates that they had too much of the genetic material of smallpox in their laboratories. Laboratories are allowed to have 20% of the full virus genome, according to international agreement.[56] Holding more smallpox DNA than is technically allowed doesn't *necessarily* pose a safety risk—but it does demonstrate an alarming lack of familiarity with biosafety and biosecurity regulations not only by the researchers involved, but also by the reviewers of their research.

Synthetic Infections

The final category of concern for synthetic biology biosafety risks is the possibility of a consequential error. Due to improper biosafety containment or scientific ignorance, a synthetic pathogen could be released and spread, causing harm to people, animals, or the environment. Theoretically, if such a laboratory-created pathogen were transmissible and novel, so that there is no existing immunity in the population, such an accident could become a global disaster. In 1918, the naturally occurring outbreak of a novel, transmissible strain of influenza killed more than 50 million people worldwide, and the natural emergence of SARS caused 774 deaths among 8,000 cases in 2003, as well as a $40 billion loss to the global economy.[57,58]

Consequential laboratory accidents have happened before, but thankfully with limited effects on human health. For example, in 2003-04, there were multiple laboratory acquired infections (LAIs) with SARS in biological research laboratories, but transmission was halted before the disease could spread widely.[59-63] A more unusual event occurred in 1977, when an influenza strain started circulating that was essentially identical to a virus that had been circulating in the 1950s.[64] It was as if it had been frozen in time—or perhaps, a frozen laboratory test tube. The unusual re-emergent virus was atypically mild for a new strain, but it still became the dominant flu strain for that season around the world. It was particularly a problem for people 26 years old or younger, who had not been exposed to the virus when it had made its previous appearance. This led to severe outbreaks at universities and military colleges. The outbreak that occurred at the US Air Force Academy was so severe—over the course of 9 days, 76% of the students, or 3,280 cadets, became ill—that all academic and military training was suspended. This was the "first such interruption in training due to influenza illness in the cadet population."[65] Given what is known about viral evolution, and how strains evolve rapidly from year to year, this flu epidemic was almost certainly not a natural event. It has often been blamed on a laboratory accident in the years since, but this is not a likely scenario.[66] The weight of the evidence, which includes genetic sequence data as well as statements made at the time, strongly suggest that this unnatural flu outbreak was the result of a poorly designed and executed vaccine trial involving thousands of Chinese military recruits.[67,68] It was not likely the result of a single laboratory containment failure.

Nonetheless, the possibility that a novel pathogen could escape containment is a serious concern and was a major reason for the controversy surrounding so-called gain-of-function (GOF) influenza research. In GOF research, pathogens are scientifically altered to introduce traits and functions not found in the wild; the research aims, as described by the US government, "to increase the ability of infectious agents to cause disease by enhancing its pathogenicity or by increasing its transmissibility."[69] The research first caused controversy in 2011, when 2 laboratories in the US and the Netherlands deliberately sought to make H5N1 avian influenza more transmissible between mammals, in order to determine how and whether this transition was likely to occur in nature and which genetic mutations would indicate that the virus was evolving to become more dangerous to humans.

The GOF influenza work that sparked the initial controversy is not technically considered to be synthetic biology, because the experiments were set up so that the virus could *evolve* into a more transmissible pathogen—that is, it was not specifically *designed* to be more transmissible. However, the research became emblematic of potential synthetic biology and advanced biotechnology concerns, including the risks of democratization of the technology.[70] While the original experiments were performed by world-class experts in world-class facilities, synthetic biology techniques could allow such influenza strains to be replicated in laboratories with less-robust safety systems, health monitoring, and extensive experience with dangerous pathogens. As trends toward democratization continue, this will mean that more and more people have the capacity to cause an epidemic with a modified flu strain, even by accident.

The importance of biosafety, generally, has been underlined after some high-profile incidents at US government facilities. In 2014, there were several such incidents; while they also did not involve synthetic biology, they are nonetheless indications that safety procedures are not what they should be and that errors can occur even in premier laboratories. At the Centers for Disease Control and Prevention (CDC), 70 people had to be placed on postexposure prophylaxis with antibiotics and vaccine after they were unwittingly exposed to improperly inactivated *Bacillus anthracis,* the causative agent of anthrax disease.[71] Again at CDC but at a different laboratory, a less harmful, low pathogenic, avian influenza strain was contaminated with a highly pathogenic strain and shipped to an agricultural research center.[72] Appropriate precautions were taken in working with the strain, even though its true pathogenicity was not known to the researchers. No inadvertent exposures resulted, and no one fell ill.

In 2014, there was also a reminder that biosafety lapses are not a new phenomenon. At the US Food and Drug Administration (FDA), decades-old glass vials were discovered that were labeled "variola," the causative agent of smallpox.[73] These vials were later found to actually contain live virus. Smallpox was declared eradicated by the World Health Organization (WHO) in 1980, and all laboratories that held samples of the virus were supposed to have destroyed them or sent them on to WHO, to be held at the only 2 laboratories allowed to keep them—the CDC and a Russian laboratory. Given that many years had passed and these samples had not been either transferred or disposed of, this incident was indicative of poor inventory management procedures that had been going on for some time. No one was exposed to the smallpox virus in the course of this incident.

Limits in scientific knowledge have also led to errors. In 2015, it was discovered that the US Army Dugway Proving Ground had shipped samples containing live anthrax to centers that were ill prepared (and not registered) to work with it. These samples were irradiated so that the anthrax spores would be killed and not cause disease, so that companies that develop detectors and other biodefense technologies could use them without risk of infection or having to invest in getting the appropriate clearances to work with regulated pathogens that are on the US select agent list. The anthrax samples were given high doses of radiation—124 kGys—and were tested to see if the anthrax spores were still able to grow. After several days of no evidence of growth, the samples were given a "death certificate" and distributed to biodefense companies and laboratories all over the world. Nonetheless, some spores "recovered"—and some shipments were determined to have live anthrax spores. The shipments involved every state in the United States and several countries.[74,75] No one became ill as a result of these incidents.

These repeated biosafety incidents have called into question the adequacy of US national policies and procedures aimed at preventing biological accidents and the adequacy of the knowledge that we have about biological systems in order to keep the larger community safe, and they have prompted US government action. There have been multiple reviews of the biosafety lapses in US government laboratories and changes in leadership and responsibility at the CDC and DoD to deal with research and biosafety oversight.[76-78] In addition, a US government funding moratorium was placed on GOF influenza research until the risks and benefits of such research could be evaluated.[69,79]

The Obama administration released a memo to US government agencies outlining a schedule for enhancing biosafety and biosecurity at infectious disease laboratories, including developing a new voluntary, anonymous, nonpunitive incident reporting system, similar in theory to the system used by the aviation industry that encourages reporting with the intent to learn quickly from safety lapses.[80] These steps have been recommended for years and should help to increase the safety of all laboratory personnel.[81]

Currently, there is a lack of information about biosafety incidents and possible exposures to pathogens in the laboratory beyond the regulated pathogens (select agents), which are heavily monitored. Without an incentive to come forward with information about incidents, or mechanisms to report anonymously, it is difficult to get people to report; any incident or laboratory acquired infection is embarrassing both for the person infected and for the institution. However, a lack of information about biosafety lapses not only prevents safety best practices from being developed and promulgated, it is also dangerous if serious biosafety lapses are not reported. If the accident and incident reporting system is modeled after the reporting that takes place in the aviation industry, operational experiences and incidents may be reported without fear of punitive action, and the CDC would analyze mistakes without attributing them to individuals, except when incidents are the result of criminal conduct.

Biosafety also needs to be studied, because there are big gaps in knowledge about what incidents and accidents occur in laboratories and how to make scientific research safer. There are behavioral studies to be done: how best to instill a safety culture in the laboratory, how to develop the best training material, how to inspect laboratories in such a way as to improve safety over time, and how to promote safe practices even in routinized biological laboratory environments. In addition, comparative studies are needed for practices, engineering, laboratory set-ups, and equipment.[82] Data gaps could be the foundation of a master's or doctoral thesis—but there is currently no such scholarship and research in this area to address those data gaps. Without specific funding of biosafety scholarship, publications that put biosafety research information into the public domain will appear only occasionally.

Development of International Norms

There is also more that can be done internationally to promote safety in the laboratory and beyond. Biosafety is often not perceived as important enough to receive resources dedicated to training, oversight, equipment, standards, and other mechanisms to protect the public's health—even in rich donor countries. There is guidance available for laboratories to develop oversight systems to catch and contain accidents, but not all research institutions adhere to such guidance, require adequate training, or have sufficient resources to dedicate to biosafety. For example, there are international standards for BSL-1, BSL-2, BSL-3, and BSL-4 labs, including what engineering controls should be in place in each level of biocontainment to manage biorisks in a research institution.[83,84] The WHO, CDC, professional organizations, and other institutions aim to bring technical information to practitioners, enhance laboratory safety practice, and promote biosafety standards.[85-89]

Yet, while technical guidance for researchers and institutions is in abundance, a key piece is missing: *national-level* norms for the safety systems necessary to perform such consequential research, to make biosafety a political priority.[90] Nations should have norms and expectations of each other that they will maintain a biosafety infrastructure that is capable of preventing and mitigating consequences from a laboratory accident. There is currently no international agreement or treaty that explicitly covers this issue of national responsibility to maintain biosafety for biological research. Even among those mechanisms that relate to biology or infectious disease control, such as the International Health Regulations (2005), Biological Weapons Convention, the Global Health Security Agenda, or the Cartagena Protocol on Biosafety, the responsibilities and international expectations to prevent a consequential laboratory accident in biological research endeavors is not adequately addressed.[91] And in a study of the biosafety regulations of 10 nations—Brazil, China, India, Israel, Pakistan, Kenya, Russia, Singapore, the United Kingdom, and the United States—there was a wide divergence in the quality and quantity of biosafety regulations covering pathogens that could result in a consequential laboratory accident. Advanced or synthetic biology were not often specifically addressed, and funding information for biosafety was generally unavailable as well.[92]

By explicitly discussing and codifying these norms in an international context, there may be more political pressure for nations to duly allocate resources to prevent catastrophic biocontainment failures. The next time there is alarm about GOF or some other potentially concerning research, it would be helpful to know that the research took place in an environment where there are national standards for the work—for example, standards for equipment maintenance, worker safety training, health monitoring, surveillance, and other myriad activities that experts believe can keep the researchers and the public safe. It would be good to know that the nation has an adequate public health surveillance and response system in place to identify and limit any outbreaks that could result from such accidents. Nations that fund this type of scientific research should therefore have the systems in place to provide appropriate levels of safety. Even the most dangerous pathogen cannot cause harm to populations if it does not escape containment.

Conclusion

Addressing the biosafety concerns in synthetic biology will require updating and expanding biosafety training programs and national policies. It will require reporting of accidents and near-misses, so that mistakes can be learned from and accidents prevented. It will require research into new methods of biosafety techniques like intrinsic biosafety, but also new ways to make laboratory procedures and equipment safer. Because no nation can protect itself from the spread of infectious diseases, it will require that nations develop international norms for how to train, monitor, fund, and ensure biosafety.

Addressing biological safety issues for synthetic biology is important for the health and safety of synthetic biology practitioners (and their close contacts), whether they are working in an academic, biotech, or community laboratory—or in their kitchen. But it is also important for realizing the promise of synthetic biology. Modernizing and being proactive about safety standards and procedures that take synthetic biology into account presents an opportunity to prepare for what comes next, so that the inevitable future challenges to safety from synthetic biology, or another yet-to-be-named biological technology, can be met.

References

1. Church G. De-extinction is a good idea. *Scientific American* September 1, 2013. http://www.scientificamerican.com/article.cfm?id=george-church-de-extinction-is-a-good-idea. Accessed August 26, 2016.
2. Braat H, Rottiers P, Hommes DW, et al. A phase I trial with transgenic bacteria expressing interleukin-10 in Crohn's disease. *Clin Gastroenterol Hepatol* 2006;4(6):754-759.
3. Caluwaerts S, Vandenbroucke K, Steidler L, et al. AG013, a mouth rinse formulation of Lactococcus lactis secreting human Trefoil Factor 1, provides a safe and efficacious therapeutic tool for treating oral mucositis. *Oral Oncol* 2010;46(7):564-570.
4. Salomon DR. A CRISPR way to block PERVs—engineering organs for transplantation. *N Engl J Med* 2016;374(11):1089-1091.
5. Cai Y, Agmon N, Choi WJ, et al. Intrinsic biocontainment: multiplex genome safeguards combine transcriptional and recombinational control of essential yeast genes. *Proc Natl Acad Sci U S A* 2015;112(6):1803-1808.
6. Benner SA. *Life, the Universe and the Scientific Method.* Gainesville, FL: FfAME Press; 2009.
7. Moe-Behrens GH, Davis R, Haynes KA. Preparing synthetic biology for the world. *Front Microbiol* 2013;4:5.
8. Schmidt M. Xenobiology: a new form of life as the ultimate biosafety tool. *Bioessays* 2010;32(4):322-331.
9. Lajoie MJ, Rovner AJ, Goodman DB, et al. Genomically recoded organisms expand biological functions. *Science* 2013;342(6156):357-360.
10. Mandell DJ, Lajoie MJ, Mee MT, et al. Biocontainment of

genetically modified organisms by synthetic protein design. *Nature* 2015;518(7537):55-60.

11. Dutchen S. Biological safety lock for genetically modified organisms. *Phys.org* January 21 2015. http://phys.org/news/2015-01-biological-safety-genetically.html. Accessed August 26, 2016.
12. Wade N. Synthetic bacterial genome takes over a cell, researchers report. *New York Times.* May 20, 2010:A17. http://www.nytimes.com/2010/05/21/science/21cell.html. Accessed August 26, 2016.
13. Grant B. News in a nutshell. *Scientist.* March 31 2011. http://www.the-scientist.com/?articles.view/articleNo/29636/title/News-in-a-nutshell/. Accessed August 26, 2016.
14. First self-replicating synthetic bacterial cell [press release]. J. Craig Venter Institute, May 20, 2010. http://www.jcvi.org/cms/press/press-releases/full-text/article/first-self-replicating-synthetic-bacterial-cell-constructed-by-j-craig-venter-institute-researcher/home/. Accessed August 26, 2016.
15. Venkataramanan M. 'Track changes' for your genes: DARPA goal. *Wired* March 29, 2011. http://www.wired.com/2011/03/track-changes-for-your-genes-darpa-goal/. Accessed August 26, 2016.
16. Dana GV, Kuiken T, Rejeski D, Snow AA. Synthetic biology: four steps to avoid a synthetic-biology disaster. *Nature* 2012;483(7387):29.
17. Holdren JP, Shelanski H, Vetter D, Goldfuss C. Improving transparency and ensuring continued safety in biotechnology. July 2, 2015. https://www.whitehouse.gov/blog/2015/07/02/improving-transparency-and-ensuring-continued-safety-biotechnology. Accessed August 26, 2016.
18. Evans A. Glowing plants: natural lighting with no electricity. Kickstarter 2013. https://www.kickstarter.com/

projects/antonyevans/glowing-plants-natural-lighting-with-no-electricit. Accessed April 1, 2016.

19. Callaway E. Glowing plants spark debate. *Nature* June 4, 2013. http://www.nature.com/news/glowing-plants-spark-debate-1.13131. Accessed August 26, 2016.
20. Regalado A. Why Kickstarter's glowing plant left backers in the dark. *MIT Technology Review* July 15, 2016. https://www.technologyreview.com/s/601884/why-kickstarters-glowing-plant-left-backers-in-the-dark/. Accessed August 26, 2016.
21. Carter SR, Rodemeyer M, Garfinkel MS, Friedman R. *Synthetic Biology and the U.S. Biotechnology Regulatory System: Challenges and Options.* J. Craig Venter Institute; May 2014. http://www.jcvi.org/cms/fileadmin/site/research/projects/synthetic-biology-and-the-us-regulatory-system/full-report.pdf. Accessed August 26, 2016.
22. Landers J. Can scientists bring mammoths back to life by cloning? *Washington Post* February 9, 2015. http://www.washingtonpost.com/national/health-science/can-scientists-bring-mammoths-back-to-life-by-cloning/2015/02/06/2a825c8c-80ae-11e4-81fd-8c4814dfa9d7_story.html. Accessed August 26, 2016.
23. Vogel G. Top mosquito suspect found infected with Zika. *Science Insider.* May 23, 2016. http://www.sciencemag.org/news/2016/05/top-mosquito-suspect-found-infected-zika. Accessed August 26, 2016.
24. Dick GW, Kitchen SF, Haddow AJ. Zika virus. I. Isolations and serological specificity. *Trans R Soc Trop Med Hyg* 1952;46(5):509-520.
25. European Centers for Disease Prevention and Control (ECDC). Zika virus epidemic in the Americas: potential association with microcephaly and Guillain-Barré syndrome. December 10, 2015. http://ecdc.europa.eu/en/publications/Publications/zika-virus-americas-association-

with-microcephaly-rapid-risk-assessment.pdf. Accessed August 26, 2016.

26. Pan American Health Organization, World Health Organization. Regional Zika epidemiological update (Americas), July 14, 2016. http://bit.ly/2da62tr. Accessed August 26, 2016.
27. McNeil DG Jr. Growing support among experts for Zika advice to delay pregnancy. *New York Times* February 5, 2016. http://www.nytimes.com/2016/02/09/health/zika-virus-women-pregnancy.html?_r=0. Accessed August 26, 2016.
28. Centers for Disease Control and Prevention. CDC issues advice for travel to the 2016 Summer Olympic Games [press release]. February 26, 2016. http://www.cdc.gov/media/releases/2016/s0226-summer-olympic-games.html. Accessed August 26, 2016.
29. McKay B, Bisserbe N. Sanofi teams up with U.S. Army on Zika vaccine. *Wall Street Journal* July 6, 2016. http://www.wsj.com/articles/sanofi-teams-up-with-u-s-army-on-zika-vaccine-1467781202. Accessed August 26, 2016.
30. Saey TH. Gene drives spread their wings. *Science News* December 2, 2015. https://www.sciencenews.org/article/gene-drives-spread-their-wings. Accessed August 26, 2016.
31. Adelman ZN, Tu Z. Control of mosquito-borne infectious diseases: sex and gene drive. *Trends Parasitol* 2016;32(3):219-229.
32. Gantz VM, Jasinskiene N, Tatarenkova O, et al. Highly efficient Cas9-mediated gene drive for population modification of the malaria vector mosquito Anopheles stephensi. *Proc Natl Acad Sci U S A* 2015;112(49):E6736-6743.
33. World Health Organization. 10 facts on malaria. Updated November 2015. http://www.who.int/features/factfiles/malaria/en/. Accessed August 26, 2016.

34. Oye KA, Esvelt K, Appleton E, et al. Biotechnology. Regulating gene drives. *Science* 2014;345(6197):626-628.
35. Regalado A. We have the technology to destroy all Zika mosquitoes. *MIT Technology Review* February 8, 2016. https://www.technologyreview.com/s/600689/we-have-the-technology-to-destroy-all-zika-mosquitoes/. Accessed August 26, 2016.
36. Pennisi E. Gene drive turns insects into malaria fighters. *Science* November 23, 2015. http://www.sciencemag.org/news/2015/11/gene-drive-turns-insects-malaria-fighters. Accessed August 26, 2016.
37. Akbari OS, Bellen HJ, Bier E, et al. Biosafety. Safeguarding gene drive experiments in the laboratory. *Science* 2015;349(6251):927-929.
38. Committee on Gene Drive Research in Non-Human Organisms; Recommendations for Responsible Conduct; Board on Life Sciences; Division on Earth and Life Studies; National Academies of Sciences, Engineering, and Medicine. *Gene Drives on the Horizon: Advancing Science, Navigating Uncertainty, and Aligning Research with Public Values.* Washington, DC: National Academies Press; 2016.
39. Pollack A. New weapon to fight Zika: the mosquito. *New York Times* January 30, 2016. http://www.nytimes.com/2016/01/31/business/new-weapon-to-fight-zika-the-mosquito.html?_r=0. Accessed August 26, 2016.
40. Carson R, Darling L, Darling L. *Silent Spring*. BostonMA: Houghton Mifflin, Riverside Press; 1962.
41. Carlson R. Splice it yourself. *Wired* May 1, 2005. http://www.wired.com/2005/05/splice-it-yourself/. Accessed August 29, 2016.
42. DIY Bio. Local groups. http://diybio.org/local/. Accessed June 2, 2016.
43. An institution for the do-it-yourself biologist. DIYBio.org web page. https://diybio.org/. Accessed August 29, 2016.
44. Zayner J. Learn modern science by aoing. 2016.

https://www.indiegogo.com/projects/diy-crispr-kits-learn-modern-science-by-doing#/. Accessed March 25, 2016.
45. Baltimore Under Ground Science Space (BUGSS). BUGSS Safety Manual. July 1, 2012. http://www.bugssonline.org/uploads/1/2/9/1/12918855/bugss_safety_manual_v_1.1.pdf. Accessed August 29, 2016.
46. Responsible science for do-it-yourself biologists new initiative launched on biosafety [news release]. Woodrow Wilson International Center for Scholars. July 29, 2010. http://www.synbioproject.org/process/assets/files/6424/diypressrelease.pdf. Accessed August 29, 2016.
47. DIY Bio. Ask a biosafety professional your question. http://ask.diybio.org/. Accessed February 22, 2016.
48. Alfred P. Sloan Foundation. Recently completed programs: synthetic biology. 2016. http://www.sloan.org/major-program-areas/recently-completed-programs/synthetic-biology/. Accessed February 22, 2016.
49. Committee on the Future of the Colleges of Agriculture in the Land Grant University System; Board on Agriculture; National Research Council. *Colleges of Agriculture at the Land Grant Universities: Public Service and Public Policy*. Washington, DC: National Academy Press; 1996.
50. iGEM: Synthetic biology based on standard parts. 2015. http://igem.org/About. Accessed April 1, 2016.
51. iGEM. iGEM 2014 team information: Heidelberg. http://igem.org/Team.cgi. Accessed May 26, 2015.
52. You E. FBI connects science and security communities. *Science, Safety, and Security Quarterly* May 2012;(2):1, 5. http://www.phe.gov/s3/Documents/s3newsletter-may2012.pdf. Accessed August 29, 2016.
53. Lempinen EW. FBI, AAAS collaborate on ambitious outreach to biotech researchers and DIY biologists. April 1, 2011. *AAAS News* http://www.aaas.org/news/fbi-aaas-

collaborate-ambitious-outreach-biotech-researchers-and-diy-biologists. Accessed August 29, 2016.

54. US Department of Health and Human Services. Screening framework guidance for providers of synthetic double-stranded DNA. *Federal Register* 2010;75(197): 62820-62832. http://www.gpo.gov/fdsys/pkg/FR-2010-10-13/html/2010-25728.htm. Accessed August 29, 2016.
55. Bluestein A. Tom Knight, godfather of synthetic biology, on how to learn something new. *Fast Company* August 28, 2012. http://www.fastcompany.com/3000760/tom-knight-godfather-synthetic-biology-how-learn-something-new. Accessed August 29, 2016.
56. Tucker JB. The smallpox destruction debate: could a grand bargain settle the issue? *Arms Control Today* March 2009. https://www.armscontrol.org/act/2009_03/tucker#Sidebar. Accessed August 29, 2016.
57. Taubenberger JK, Morens DM. 1918 influenza: the mother of all pandemics. *Emerg Infect Dis* 2006;12(1):15-22.
58. Lee J-W, McKibbin WJ. Estimating the global economic costs of SARS. In: Knobler S, Mahmoud A, Lemon S, et al, eds., *Learning from SARS: Preparing for the Next Disease Outbreak: Workshop Summary*. Washington, DC: National Academies Press; 2004.
59. Lim W, Ng KC, Tsang DN. Laboratory containment of SARS virus. *Ann Acad Med Singapore* 2006;35(5):354-360.
60. Normile D. Infectious diseases. SARS experts want labs to improve safety practices. *Science* 2003;302(5642):31.
61. Normile D. Infectious diseases. Mounting lab accidents raise SARS fears. *Science* 2004;304(5671):659-661.
62. Normile D. Infectious diseases. Second lab accident fuels fears about SARS. *Science* 2004;303(5654):26.
63. Senior K. Recent Singapore SARS case a laboratory accident. *Lancet Infect Dis* 2003;3(11):679.
64. Zakstelskaja LJ, Yakhno MA, Isacenko VA, et al. Influenza

in the USSR in 1977: recurrence of influenzavirus A subtype H1N1. *Bull World Health Organ* 1978;56(6):919-922.
65. US Air Force. Epidemiologic investigation of A/USSR/90/70(H1N1) at the US Air Force Academy, Colorado, 3 to 13 February 1978. 1978:1-10.
66. Doshi P. Trends in recorded influenza mortality: United States, 1900-2004. *Am J Public Health* 2008;98(5):939-945.
67. Rozo M, Gronvall GK. The reemergent 1977 H1N1 strain and the gain-of-function debate. *MBio* 2015;6(4).
68. Palese P. Influenza: old and new threats. *Nat Med* 2004;10(12 Suppl):S82-87.
69. White House Office of Science and Technology Policy. Doing diligence to assess the risks and benefits of life sciences gain-of-function research. October 17, 2014. http://www.whitehouse.gov/blog/2014/10/17/doing-diligence-assess-risks-and-benefits-life-sciences-gain-function-research. Accessed August 29, 2016.
70. Gronvall GK. H5N1: a case study for dual-use research. Council on Foreign Relations Working Paper. July 2013. http://www.cfr.org/public-health-threats-and-pandemics/h5n1-case-study-dual-use-research/p30711. Accessed August 29, 2016.
71. CDC lab determines possible anthrax exposures: staff provided antibiotics/monitoring [perss release]. June 19, 2014. Centers for Disease Control and Prevention. http://www.cdc.gov/media/releases/2014/s0619-anthrax.html. Accessed August 29, 2016.
72. Center for Disease Control and Prevention. Summary of the inadvertent shipment of an influenza virus H5N1-containing laboratory specimen. July 11, 2014. http://www.cdc.gov/flu/news/h5n1-influenza-shipment.htm. Accessed August 29, 2016.
73. Update on findings in the FDA cold storage area on the NIH campus [press release]. July 16, 2014. US Food and Drug Administration. http://www.fda.gov/NewsEvents/

Newsroom/PressAnnouncements/ucm405434.htm. Accessed August 29, 2016.

74. U.S. Department of Defense. Department of Defense laboratory review. 2015. http://www.defense.gov/News/Special-Reports/DoD-Laboratory-Review. Accessed April 1, 2016.
75. Vanden Brook T, Young A. Pentagon's anthrax scandal spreads to Canada. *USA Today* June 3, 2015. http://www.usatoday.com/story/news/2015/06/01/anthrax-pentagon-scandal/28329125/. Accessed August 29, 2016.
76. *Report of the Federal Experts Security Advisory Panel.* 2014. http://www.phe.gov/s3/Documents/fesap.pdf. Accessed August 29, 2016.
77. National Science and Technology Council, Committee on Homeland and National Security, Subcommittee on Biological Defense Research and Development, Fast Track Action Committee on the Select Agents Regulations. *Fast Track Action Committee Report: Recommendations on the Select Agent Regulations Based on Broad Stakeholder Engagement.* October 2015. http://www.phe.gov/s3/Documents/ftac-sar.pdf. Accessed August 29, 2016.
78. Young A. Top U.S. lab regulator replaced in wake of incidents with bioterror pathogens. *USA Today* December 8, 2015. http://www.usatoday.com/story/news/2015/12/08/cdc-bioterror-lab-regulator-replaced/76976554/. Accessed August 29, 2016.
79. Gryphon Scientific. *Risk and Benefit Analysis of Gain of Function Research.* December 2015. http://www.gryphonscientific.com/wp-content/uploads/2015/12/Final-Gain-of-Function-Risk-Benefit-Analysis-Report-12.14.2015.pdf. Accessed August 29, 2016.
80. *Implementation of Recommendations of the Federal Experts Security Advisory Panel (FESAP) and the Fast Track Action Committee on Select Agent Regulations (FTAC-SAR).* October

2015. http://www.phe.gov/s3/Documents/fesap-ftac-ip.pdf. Accessed August 29, 2016.
81. Testimony of Gigi Kwik Gronvall, Hearing on Germs, Viruses, and Secrets: The Silent Proliferation of Bio-Laboratories in the United States. US House of Representatives Committee on Energy and Commerce, Subcommittee on Oversight and Investigations. October 4, 2007. http://www.upmchealthsecurity.org/our-work/testimony/hearing-on-germs-viruses-and-secrets. Accessed August 29, 2016.
82. Gronvall GK, Shearer MP, Collins H, Inglesby T. *Improving Security Through International Biosafety Norms.* July 2016. UPMC Center for Health Security. http://www.upmchealthsecurity.org/our-work/pubs_archive/pubs-pdfs/2016/Final_report_to_PASCC_071416.pdf. Accessed August 29, 2016.
83. European Biosafety Association. *Update on Future of CWA 15793 Laboratory Biorisk Management.* 2014. http://dechema.de/ebsa/cwa_15793.html. Accessed August 29, 2016.
84. CEN Workshop Agreement. *CWA 15793.* Brussels: European Committee on Standardization; 2011. http://www.iso.org/sites/biotechnology2011/resources.html. Accessed August 29, 2016.
85. World Health Organization. *Laboratory Biosafety Manual.* 3d ed. Geneva: World Health Organization; 2004. http://www.who.int/csr/resources/publications/biosafety/en/Biosafety7.pdf. Accessed August 29, 2016.
86. Centers for Disease Control and Prevention. *Biosafety in Microbiological and Biomedical Laboratories.* 5th ed. Washington, DC: U.S. Department of Health and Human Services, Public Health Service, Centers for Disease Control and Prevention, National Institutes of Health; 2009.

http://www.cdc.gov/biosafety/publications/bmbl5/BMBL.pdf. Accessed August 29, 2016.

87. American Biological Safety Association. 2016. http://www.absa.org/. Accessed April 1, 2016.
88. European Biosafety Association. 2016. www.ebsaweb.eu. Accessed April 1, 2016.
89. Asia-Pacific Biosafety Association. www.a-pba.org. Accessed April 1, 2016.
90. Gronvall GK, Rozo M. Addressing the gap in international norms for biosafety. *Trends Microbiol* 2015;23(12):743-744.
91. Gronvall GK, Rozo M. *Synopsis of Biological Safety and Security Arrangements.* UPMC Center for Health Security. July 20, 2015. http://bit.ly/2dnZdHk. Accessed August 29, 2016.
92. Gronvall GK, Shearer MP, Collins H. *National Biosafety Systems.* UPMC Center for Health Security. July 2016. http://bit.ly/2dDJiBG. Accessed August 29, 2016.

CHAPTER 4.

ON ETHICS AND PUBLIC ENGAGEMENT

The most persistent concern about morality and ethics in the biological sciences is whether scientists are "playing God."[1,2] Given the extraordinary power of synthetic biology research tools, such accusations of scientific hubris and overreach may be inevitable. Indeed, one research objective in the synthetic biology field aims to mimic and better understand the conditions that sparked life on earth—in other words, these synthetic biologists aim to *literally* create life. Far removed from the plot lines of Mary Shelley's Frankenstein, these scientists study protocells, which are self-organized lipid structures, because these simple structures are thought to be an intermediate form between nonliving material and cellular life.[3-5]

Synthetic biology projects have generated multiple newsworthy controversies that invoked ethics concerns and a diversity of opinions about the scientific value of the work. Each of these examples of synthetic biology research has led to examinations, some ongoing, of whether the work should have been done, whether the rules guiding the work are sufficient, and what should happen next. Some controversies were generated by synthesis projects, such as the laboratory synthesis of infectious polio virus in 2002, the recreation of the 1918 pandemic influenza virus, and the synthesis of an entire bacterial cell.[6-8] Other hot-button issues included the laboratory development of an H5N1 avian influenza ("bird flu") virus capable of being transmitted between mammals, as well as a crowd-funding campaign to produce a plant that glows in the dark.[9-11] With the recent development of CRISPR, a powerful new synthetic biology tool that permits genomes to be edited like a Word document, the possibilities for synthetic biology have exploded along with public discussions of bioethics: CRISPR could be used to modify the human germline to edit out hereditary diseases, to reduce the population of mosquitoes that carry diseases like Zika or malaria, and to "'brew" opiates from yeast.[12-19]

These past cases of controversy in synthetic biology are all different; some have led to additional guidance for scientists and new regulations, and some have not. Yet, what has undoubtedly helped to resolve the controversies is the presence of a community of people who know what regulations and governance are already in place, who can debate the scientific and ethical issues, and who can engage with the public with scientifically valid information. Ideally, when an ethical dilemma arises in synthetic biology, a balance should be struck to safeguard the potential for beneficial future discoveries and applications, but also to set safety and ethics standards for synthetic biologists to follow in the present. This balance can benefit the public both by keeping ethics considerations in the forefront as well as by letting valuable science continue. The presence of an active, informed community has helped to get closer to achieving this balance and also helped to promote the correct order of actions. For example, discussions of whether it is ethical to proceed with a scientific project should come before the work is attempted, not after: Just because something *can* be done doesn't necessarily mean that it *should* be done.

Communities of people who are well informed about ethics, guidance, and regulations, including scientists, ethicists, and the interested public, are not formed by accident. They have to be cultivated and maintained. This chapter describes how such communities have been created in the past to grapple with synthetic biology's ethical concerns and what should be done to protect and guide the field so that it can yield benefits to the public. While building a community of interest is an ever-present education need that requires sustained investment, there may be no other way to deal with the social and ethical implications of new synthetic biology so that the benefits of the technology to society can be realized and the risks reduced.

The Beginning of Synthetic Life

A major event in the ethical considerations of synthetic biology occurred on May 20, 2010. Craig Venter, one of the first scientists to sequence the human genome, announced that he and his team had created *Mycoplasma mycoides JCVI-syn1.0*, "the first self-replicating species that we have had on the planet whose parent is a computer."[20] At more than a million base pairs of DNA, the genome of the synthetic cell was also the largest to be chemically synthesized, stitched together, and booted up like a computer software program in order to control a living, replicating cell.[20] In case there was any doubt about the not-natural origin of the strain, the scientists inserted coded messages, or watermarks, into its genome, including the names of the cell designers, the species' own email address, and a quote from physicist Richard Feynman that professed the aspirations of the scientists: "What I cannot build, I cannot understand."[21]

Building a synthetic cell was not a trivial undertaking. The project took more than 15 years for 25 scientists at the J. Craig Venter Institute (JCVI), at a cost of $40 million.[22] Three months of work were lost trying to figure out why the genome wouldn't boot up when it was expected to—and the culprit turned out to be 1 incorrect DNA base pair out of a million, the tiniest of needles in a massive haystack.[23] The technical complexity of the project ended up being much, much more than the scientists expected. When work began in 1995, the synthesis of the genome was supposed to be just the first step, so that the scientists could move on to their real goal: to experimentally determine the fewest number of genes, or the minimal genome, that could sustain life.[24] Once the scientists built an organism stripped of all excess genetic material, they could use it as an industrial workhorse, putting in new genes so that the cell could produce industrially important compounds.[25] Ultimately, the difficulty of the first step of that project left them, 15 years later, at the beginning of that quest, but with a proof of concept in hand to manufacture more novel synthetic organisms.

Further work by the JCVI scientists to build the minimal genome was still planned as a next step; the work would not be applied to the bacterial synthetic cell but to eukaryotic algae. At the press conference, Venter said they would build "an entire algae genome so we can vary the 50 to 60 different parameters for algae growth to make superproductive organisms."[26] Exxon was to fund the effort, with a $600 million contract with Venter's for-profit company, Synthetic Genomics, to develop the to-be-engineered algae to produce gasoline and diesel fuel.[27] They also planned to design cells to produce "new food oils, and new biological derived sources of plastic and chemicals" as well as new "antibiotic compounds that are currently too complex for chemists to make."[27] Venter saw the synthetic cell bringing on a future that allows manufacturing to be greener, cheaper, and distributed so that important products can be made close to where they are needed.

President Obama and Pope Benedict XVI called Venter on the day of the synthetic cell announcement.[22] Venter had been conspicuously careful not to say that they had created life, as the synthesized genome had been transplanted into an existing cell, and over rounds of replication, the original cell materials were replaced by components made from the synthetic genome.[28] But there was an immediate sense that what the scientists did in creating *Mycoplasma mycoides JCVI-syn1.0* held ethical and societal implications. Accusations that the scientists were "playing God" flooded the press.[29,30] On the same day as the announcement, President Obama charged the newly appointed Presidential Commission on the Study of Bioethical Issues to take synthetic biology on as their first order of business, and the US House Energy and Commerce Committee called for hearings about the synthetic cell.[31,32]

The reaction from the technical community was mixed. Some scientists hailed the synthetic cell as a significant achievement, but many decried the work as "glitzy," saying the import had been "overplayed."[20,26] Some questioned whether the genome was truly synthetic, as the synthetic organism's DNA was an almost-identical copy of the original bacterial strain and was thus not a completely redesigned life form with genes that didn't already exist in nature. Technical observers were also not surprised by the announcement. Almost all science is incremental, with slices of discovery written up in thousands of scientific journals accessible all over the world. The Venter work was no exception, as it built on a legacy of scientific accomplishments made by his team and others, which added to some scientists' opinions that the synthetic cell was not as novel as claimed.

Scientific merit and claims aside, more than a few scientists were concerned that Venter's announcement would trigger public concern and—in their eyes—inappropriate regulation. The interest from the president and congress seemed to confirm that regulations would be coming. One scientist wrote that "media reports hyping this as a significant, alarming step forward in the creation of artificial forms of life can be discounted," and another said, referring to Venter, "The only regulation we need is of my colleague's mouth."[22,33]

The synthetic cell work was not conducted in a regulation vacuum; there are rules, regulations, laws, and expectations that provided checks on the safety and security of the work and governed the conduct of the research. The ethics of undertaking genetic engineering have also been well explored in hundreds of books, reports, and scholarly articles. The question, however, was whether all of those rules, regulations, and considerations of the ethics of pursuing this type of synthetic biology were enough, or if the risks posed by this type of research were novel, requiring new mechanisms to evaluate the ethics of pursuing the research and new regulations for the work.

Building a Synthetic Biology Ethics Community

That synthetic biology might have significant ethical and social implications has been known since the field's beginnings.[34] The technologies are powerful, accessible, and personal and can be used in ways that may affect the lives of many on a scale not imagined before. George Church, synthetic biologist and professor at Harvard, wrote, "Any technology that can accomplish such feats—taking us back into a primeval era when mammoths and Neanderthals roamed the earth—is one of unprecedented power. Genomic technologies will permit us to replay scenes from our evolutionary past and take evolution to places where it has never gone, and where it would probably never go if left to its own devices."[35(pp11-12)]

In spite of the fears of some scientists about additional regulatory burdens, synthetic biologists have largely been receptive to discussions of ethics and public engagement in the work. Before embarking on the minimal genome synthetic cell project, the JCVI scientists first assembled a group of ethicists to examine whether the project should go forward, in work published in *Science* in 1999.[36] It was an esteemed group, including Arthur

L. Caplan, a well-known bioethicist who has been hailed for translating philosophical debates into understandable ideas and for helping the public develop informed opinions about biotechnology.[37] Though the ability to create an organism with a minimal genome was at that time "beyond current technology," the group of ethicists identified ethical, social, and religious issues raised by the research as well as the potential abuses of the technology and ultimately concluded that no major ethical objections should prevent the work from going forward.[36]

Receptivity of synthetic biologists to engagement with social scientists, ethicists, and policy experts was seen as an opportunity to guide the field to promote the public interest. Some funding to research the ethical, legal, and social implications (ELSI) for synthetic biology came as a percentage of funding for scientific work. For example, Synberc, a multi-university research center established in 2006 through a grant from the National Science Foundation to lay the foundation for synthetic biology, had as part of its mission engaging the public about the challenges and opportunities of synthetic biology. They have had social scientists on site to collaborate and deepen understanding about what new ethical and governance challenges synthetic biology might bring about.[38,39] UK funding for synthetic biology ethics and governance followed a similar path, wherein social implications were funded along with the research.[40]

A great deal of the progress in understanding of the ethical challenges, governance, and social risks and benefits of synthetic biology, however, was due to the interest of a private foundation. An entire research agenda was launched with the Alfred P. Sloan Foundation's $10 million investment in synthetic biology, starting in 2005, which formed a community actively investigating the ethics, governance, and public policy issues of synthetic biology.[41] Law enforcement objectives, through partnerships of scientific organizations with the FBI, were also facilitated by the foundation.[42]

Most of the Sloan effort was concentrated on bringing respected research institutions into the field, including the Hastings Center for Bioethics, the Woodrow Wilson International Center for Scholars, the National Academies, and the J. Craig Venter Institute. The Hastings Center, founded in 1969, is the world's first bioethics institute and a highly respected voice on ethical issues related to biological science and medicine. A defining contribution of their synthetic biology work was that the field does not require the formation of a new type of bioethics. In a *Science* article, researcher Erik Parens and colleagues wrote that the questions that synthetic biology raises are familiar and related to other biotechnological concerns, and that "although creating such a [synthetic bio-ethics] subfield might be in the short-term self-interest of bioethicists, in the long run, further balkanization of bioethics would be a mistake."[43(p1449)] From there, Hastings Center researchers mapped relevant potential physical and nonphysical harms—such as injuries to fairness and equality—that could result from applications of synthetic biology and other emergent technologies.[44] Their work identified and evaluated the possible good and bad consequences of synthetic biology applications for human welfare and developed a set of general moral considerations to inform public discourse and policy on synthetic biology.[45]

The Wilson Center's work took a broader approach to studying the societal and ethical implications of synthetic biology. They aimed to improve understanding of evolving public perceptions of synthetic biology, clarify whether existing systems of regulation and regulatory institutions could address risks associated with synthetic biology, and inform and educate policymakers. The project was led by David Rejeski, who had previously performed comparable research into the ethical and societal impacts of nanotechnology. Early on, in 2008-09, the

Wilson Center mapped out the policy issues relevant to synthetic biology and established a website that became a resource for the synthetic biology field.[46] The Wilson Center project team conducted opinion surveys and focus groups to understand public perceptions of synthetic biology, and they organized public and webcasted events that built an active synthetic biology community in Washington, DC. The Wilson Center also tackled issues surrounding the DIY ("do-it yourself") Bio movement, working to explain the DIY Bio and citizen science community to policymakers and to minimize biosafety and biosecurity risks through sharing of best practices and adoption of ethical codes of conduct.[47]

The National Academy of Sciences (NAS) organized a Forum on Synthetic Biology, which engaged members of the federal government, industry, academia, philanthropy, and civil society in a series of dialogues on the emerging ethical, societal, and legal issues emerging in the field.[48] The National Academies also brought the discussion of synthetic biology implications to an international level, holding several international symposia on synthetic biology with the national academies of the UK and China.[49] They also organized additional workshops that drew an international audience, including a public workshop on influenza research and a workshop for experts to examine the communication challenges associated with conducting dual-use research of concern.[50,51]

The J. Craig Venter Institute, in addition to its scientific work, has a policy center that has focused on societal issues associated with synthetic genomics. In response to the concern that people could synthesize pathogens like smallpox using synthetic biology techniques, they published *Options for Governance* in 2007, which weighed various options for regulating gene synthesis and described challenges in regulating DNA synthesis.[52] The sequence screening developed through self-governance eventually became official guidance of the US Department of

Health and Human Services.[53,54] In addition to addressing the misuse of DNA synthesis, JCVI held numerous workshops about the environmental and technological risks of synthetic biology, theological and philosophical concerns associated with synthetic biology applications, and biosafety and biocontainment challenges in synthetic biology.[55]

The collective thinking—and funding—about how synthetic biology could grow in the public's interest created a body of scholarship and a well-informed community. By 2010, when the Presidential Commission for the Consideration of Ethical Issues took on the subject of synthetic biology in order to make recommendations for US government funding and involvement, there was already a great deal of groundwork for them to consider and experts to hear from, all of whom contributed to the commission's analysis.[56] Not everyone and every project funded through the Sloan efforts and other funders were in agreement about the right path forward for synthetic biology, how best to engage the public, and what oversight was needed for the field. But as a result of the prodigious scholarship that was already under way, the safety, security, ethics, and public policy considerations related to synthetic biology were well-described in the academic and scholarly literature, as were the potential benefits of the technology. The projects yielded new information that was useful to policymakers to promote benefits and reduce risks.

Presidential Commission

The Presidential Commission for the Study of Bioethical Issues, composed of 13 scientists, ethicists, and public policy experts, was tasked to examine not only the particular milestone that the JCVI researchers had attained in order to study its implications, but to look more broadly at other synthetic biology advances and their governance. They were asked to consider the potential medical, environmental, security, and other benefits of this field of research, as well as risks to health and security. After 6 months of information gathering and deliberation, the commission was to recommend actions the federal government should take so that "America reaps the benefits of this developing field of science while identifying appropriate ethical boundaries and minimizing identified risks."[57] Over the course of 3 public meetings in Washington, DC, Philadelphia, and Atlanta, the commission heard from more than 30 witnesses and members of the public about the benefits and risks of synthetic biology. Many of the researchers who were asked to testify to the commission had already been performing work in the area and so were well-prepared to give the commission the information it needed to make a decision.

The commission's final report, *New Directions: The Ethics of Synthetic Biology and Emerging Technologies,* was released in December 2010.[56] Dr. Amy Gutmann, the chair of the commission and president of the University of Pennsylvania, stated, "We considered an array of approaches to regulation—from allowing unfettered freedom with minimal oversight and another to prohibiting experiments until they can be ruled completely safe beyond a reasonable doubt. We chose a middle course to maximize public benefits while also safeguarding against risks."[58] Ultimately, the commission issued a number of recommendations but concluded that no new regulations for synthetic biology were needed at the time.[59]

The commission called their synthetic biology strategy "prudent vigilance," and they recommended an iterative process to review risks and benefits and to revisit decisions "as warranted by additional information about risks and potential benefits."[57] They recommended the adoption of a framework for continual monitoring of the ethical and regulatory challenges that the work may pose, and, as many similar issues could be raised by other biotechnologies, they suggested that their framework could be applied more broadly.[57]

The ethical framework they suggested for approaching synthetic biology included several principal components. The first was public beneficence, to maximize the benefits to the public and minimize harm. For example, this could be through the applications of the technology to improved medical care or more ecologically friendly fuel, as well as the potential to increase economic opportunities.[57] While benefits and harms to individuals should be considered, the commission recommended that policymakers focus assessments on a societal level.

The second component was responsible stewardship, so that people may act for the betterment of all and consider future benefits and risks to populations that may result from decisions made now. This is difficult to do with emerging technologies, because understanding of the potential benefits and risks of any give technology is incomplete and uncertain. The commission sought for balance between those who, in the face of this uncertainty, may wish for a more precautionary approach until all possible risks are known, and others who are not willing to halt technological progress even in the face of potential risks.

The third component was intellectual freedom and responsibility to expand the boundaries of human exploration and achievement, as long as that freedom is pursued in "morally responsible ways."[57] The commission also endorsed "regulatory parsimony," recommending "only as much oversight as is truly necessary to ensure justice, fairness, security and safety while pursuing the public good."[57] For new technologies, there is a temptation to stifle innovation on the basis of fear of the unknown, but regulatory burdens can be counterproductive to security and safety: "without sufficient freedom to operate, tomorrow's achievements may render moot the risks of today."[57]

The track record for technological foresight, both good and bad, has not been overwhelmingly accurate.

The fourth component was democratic deliberation, including collaborative decision making, which could be revisited over time as societal risks and benefits become more known. Finally, the framework included justice and fairness, so that the benefits and burdens of synthetic biology and other emerging biotechnologies should be fairly distributed across society.

The ethics commission report was received favorably by synthetic biologists and industry organizations, but not everyone agreed with the middle-of-the-road approach taken by the presidential commission. A letter was sent to the commission chair by 58 organizations in 22 countries, including multiple member organizations of Friends of the Earth and other anti-GMO groups.[60] Their objections were based on the commission's disregard of the precautionary principle, which would dictate that there should be a moratorium on the release and commercial use of synthetic organisms until there is "a thorough study of all the environmental and socio-economic impacts of this emerging technology... [and] ...until extensive public participation and democratic deliberation have occurred on the use and oversight of this technology."[60] The civil society organizations also objected to the commission's approach to the regulation of synthetic biology, stating that a reliance on self-regulation is basically no regulation at all.

One particular group has been very active in their opposition to synthetic biology: the ETC group, an activist organization based in Canada. In addition to their calls for a complete moratorium on what they term "extreme genetic engineering," they have major objections to synthetic biology because of the harm that it may do to the small farmers who currently grow the crops that are being "replaced" by synthetic biology, including artemisinin for malaria treatment and synthetic vanilla and saffron.[61] The

ETC group's Jim Thomas has written that he was less concerned about the scientists "playing God" than about the "practical ethics of scientists-turned-entrepreneurs' 'playing business' or rather 'playing the market.' "[62]

Asilomar

Often, when there are new, potentially concerning advances in science with ethical or social implications, there are calls for scientists to put together a "new Asilomar," replicating the 1975 conference that set up the governance of the field of recombinant DNA technology, which can be considered to be the great-grandparent of the synthetic biology field. At the beginning of the DNA recombinant era, which ushered in the possibility of genetic engineering and synthetic biology, there was a great deal of uncertainty about the safety of the work. In a letter published in *Science* in 1974, leading scientists and Nobel laureates recommended that certain types of recombinant DNA experiments—those with toxins, oncogenic viruses, and antibiotic resistance—be off-limits until their safety could be evaluated and assessed in a conference held a year later.[63] That conference was held at Asilomar, California, in February 1975 and was attended by scientists, government officials, and members of the press. Eventually, after further information was generated, there was a lifting of the moratorium in 1976, as well as the creation of a new regulatory system for recombinant DNA work funded by the US government. Asilomar has become the template for scientists' responses to scientific discoveries with social and ethical implications and a symbol of the scientific community's capacity to self-govern.[64]

However, the backdrop of biotechnology in the 21st century is quite different from the 1970s in many ways: the global pervasiveness of biotechnologies and expansion of biological sciences fields; the proliferation of more jaundiced views about regulation; a sharp increase in distrust of science by the public; increased commercialization of science; and the rapid-fire newsworthy advances in science with ethical and societal implications. Also missing was a way to meaningfully include the public in the discussion, to have a say in whether the work proceeds or not, to have a say about whether tax dollars are put toward the work, or to have an oversight role so that the work continues in the public's interest.

The legendary social scientist Paul Rabinow noted that the "exclusion of the public is no longer even imaginable in the age of the internet. Gentleman's agreements of a kind that were common in 1975 are no longer imaginable, given the rise of patenting in the biotech industry. Assurances by patriarchs that safety issues can be handled through expertise and containment are no longer plausible given the global conditions of security."[65(p1110)] There is the risk that without a deliberate plan of action, people will self-organize and form a "popular technology assessment," which may not accurately reflect the benefits and risks and therefore could be less advantageous for synthetic biology and its applications.[66]

How to include the public in the process of technological examination of synthetic biology is still a challenge. The Presidential Commission on the Study of Bioethical Issues called for a continual public dialogue while the science evolves, and it worked to model such public dialogue in the course of its work. The commission held meetings that were open to the public and invited dozens of experts to testify. There were opportunities for public comments, and written comments from the public were also sent in to the commission.[57] While these mechanisms can draw some public interest and comment, there are critics of this way of engaging the public, particularly for technologically challenging topics like synthetic biology. If public engagement were truly wished for, a more intensive strategy would need to be used than has been employed so far in discussions about synthetic biology—or indeed many other examples of potential dual-use technologies.

There is a continuum of public engagement strategies, ranging from the simple to the time-intensive.[67] The first is simple *communication* of information—such as from press releases about what the National Academies will be doing about a controversial topic such as germline editing. The second type of public engagement is *consultation,* wherein the public is polled for their opinions or asked to comment about recommendations. Surveys, focus groups, and public dialogues about synthetic biology to date tend to show a low level of public knowledge about it.[68] Finally, there is *collaboration,* which is a 2-way flow of information between the public and authorities and is iterative. This type of engagement with the public is difficult and expensive, as well as time-consuming; from the perspective of the scientists involved in synthetic biology, it is also a process fraught with risks.

The risks, as perceived by the scientists, are that the technology will be banned outright or made much more difficult and that research will be halted—for example, that the ETC group and other groups with similar views will sway the public and policymakers. Of course, scientists have given the public many reasons throughout history to distrust their intentions. But scientists also have some good reasons to be concerned about bringing in "the public"; public decision making on scientific matters has resulted in some poor or short-sighted decisions, including for vaccination policies, genetically modified organisms, stem cell research, and reproductive health. Regulations have also been put into place for a variety of biological science activities that are not always commensurate with real risks and that appear to be in place solely to mollify public concerns. In fact, the *threat* of regulation of scientific practice has demonstrably helped to spur scientific self-governance measures, such the current ban on germline editing or even the Asilomar Conference in 1975.

What then should be the right course of action for synthetic biology in engaging the public, so that the benefits of synthetic biology can be realized? There is research that suggests that engaging with the public in an intensive way ultimately provides the best results, because such methods have been demonstrated to improve the quality of decision making, enhance the legitimacy of the decisions, and build capacity for future policymaking regarding similar issues that may arise in the future.[69] While the study that concluded this was focused on environmental assessment, not on biotechnologies, the questions that might be asked of the science do not need to be technical and may yield better policies. In areas where there is a great deal of uncertainty, such as the future applications of synthetic biology, having the public brought in as a partner in policy decisions could support the best future goals of synthetic biology.

Maintaining a Community

New controversial issues will continue to arise in synthetic biology that will require informed debate, including the proposed synthesis of the human genome, or de-extinction plans to bring back the wooly mammoth. There are already too many dilemmas, arising too quickly, to devote the same level of attention afforded by the Presidential Commission on the Study of Bioethical Issues. Some new events in synthetic biology have been given special consideration by scientists and policymakers, such as the ban on germline mutations using CRISPR, for instance, as well as the debates surrounding gain-of-function research. But to manage the plethora of social and ethical implications of synthetic biology, it would be helpful to be better prepared to take them on, to have the same substrate that was available to the commission when they examined the risks and benefits of synthetic biology in 2010: That is, there should be a community of knowledgeable, informed people who are able to articulate the risks and benefits of a course of action, whether it is to fund, monitor, regulate, or ban a particular synthetic biology application. This community requires scientists, as they are intimately familiar with the details of the work, but this community should not be limited to scientists.

Communities of well-informed people who can debate these issues, can generate data on social impacts, can work with synthetic biologists to ensure that the public's interest is incorporated into study designs and approaches, and can engage the public with scientifically valid information and talk in public forums need to be developed and supported. Unfortunately, the Sloan Synthetic Biology Project has drawn to a close. Their mission was to seed an emerging field, not to be its sole support for the long term. There are other foundations that have taken on specific synthetic biology issues, such as the potential to edit the human genome. Yet, it is unclear whether the current funding sources are enough to sustain the community and continue public outreach, so that the synthetic biology field can continue to progress in ways that are in the public interest and to recommend actions for policymakers to adopt so that the positive benefits of the technology are maximized.[70]

New, creative ways to build literacy and sustain an informed, involved community need to be developed. One venture that was started through Sloan funding but now has additional sponsors is the Synthetic Biology Leadership Excellence Accelerator Program (LEAP). This program aims to "create bold new visions and strategies for developing biotechnology in the public interest."[71] Every year, a group of about 20 professionals are selected for a week-long retreat, aimed at promoting knowledge about synthetic biology as well as responsible leadership. The participants come from academia, industry, and government, with expertise spanning the biosciences, law, policy, engineering, and informatics, and participants are expected to emerge from the program with a greater understanding of the policy and governance challenges facing synthetic biology and to develop professional connections to resolve them. The synthetic biology field is ripe for these types of new investments and approaches in community building, including creative programs like LEAP and other fellowship programs, to prepare to take on emerging ethical and technical challenges. While ethical dilemmas in synthetic biology may not be easily predicted, they are certain to arise, and the future benefits of synthetic biology may hang in the balance.

References

1. Dabrock P. Playing God? Synthetic biology as a theological and ethical challenge. *Syst Synth Biol* 2009;3(1-4):47-54.
2. Dragojlovic N, Einsiedel E. Playing God or just unnatural? Religious beliefs and approval of synthetic biology. *Public Underst Sci* 2013;22(7):869-885.
3. Bedau M, Parke EC. *The Ethics of Protocells: Moral and Social Implications of Creating Life in the Laboratory*. Cambridge, MA: MIT Press; 2009.
4. Deplazes A, Huppenbauer M. Synthetic organisms and living machines: positioning the products of synthetic biology at the borderline between living and non-living matter. *Syst Synth Biol* 2009;3(1-4):55-63.
5. Regis E. *What Is Life?: Investigating the Nature of Life in the Age of Synthetic Biology*. New York: Farrar, Straus and Giroux; 2008.
6. Cello J, Paul AV, Wimmer E. Chemical synthesis of poliovirus cDNA: generation of infectious virus in the absence of natural template. *Science* 2002;297(5583):1016-1018.

7. Tumpey TM, Basler CF, Aguilar PV, et al. Characterization of the reconstructed 1918 Spanish influenza pandemic virus. *Science* 2005;310(5745):77-80.
8. Gibson DG, Glass JI, Lartigue C, et al. Creation of a bacterial cell controlled by a chemically synthesized genome. *Science* 2010;329(5987):52-56.
9. Russell CA, Fonville JM, Brown AE, et al. The potential for respiratory droplet-transmissible A/H5N1 influenza virus to evolve in a mammalian host. *Science* 2012;336(6088):1541-1547.
10. Imai M, Watanabe T, Hatta M, et al. Experimental adaptation of an influenza H5 HA confers respiratory droplet transmission to a reassortant H5 HA/H1N1 virus in ferrets. *Nature* 2012;486(7403):420-428.
11. Evans A. Glowing plants: natural lighting with no electricity. 2013. https://www.kickstarter.com/projects/antonyevans/glowing-plants-natural-lighting-with-no-electricit. Accessed April 1, 2016.
12. Baltimore D, Berg P, Botchan M, et al. Biotechnology. A prudent path forward for genomic engineering and germline gene modification. *Science* 2015;348(6230):36-38.
13. Kang X, He W, Huang Y, et al. Introducing precise genetic modifications into human 3PN embryos by CRISPR/Cas-mediated genome editing. *J Assist Reprod Genet* 2016:33(5):581-588.
14. Liang P, Xu Y, Zhang X, et al. CRISPR/Cas9-mediated gene editing in human tripronuclear zygotes. *Protein Cell* 2015;6(5):363-372.
15. White House Office of Science and Technology Policy. Doing diligence to assess the risks and benefits of life sciences gain-of-function research [blog]. October 17, 2014. http://www.whitehouse.gov/blog/2014/10/17/doing-diligence-assess-risks-and-benefits-life-sciences-gain-function-research. Accessed September 1, 2016.
16. Pollack A. New weapon to fight Zika: the mosquito. *New*

York Times January 30, 2016. http://www.nytimes.com/2016/01/31/business/new-weapon-to-fight-zika-the-mosquito.html?_r=0. Accessed September 1, 2016.

17. Regalado A. We have the technology to destroy all Zika mosquitoes. *MIT Technology Review* February 8, 2016. https://www.technologyreview.com/s/600689/we-have-the-technology-to-destroy-all-zika-mosquitoes/. Accessed September 1, 2016.
18. Oye KA, Lawson JC, Bubela T. Drugs: regulate 'home-brew' opiates. *Nature* 2015;521(7552):281-283.
19. Service RF. Modified yeast produce opiates from sugar. *Science* 2015;349(6249):677-677.
20. Wade N. Synthetic bacterial genome takes over a cell, researchers report. *New York Times* May 20, 2010:A17. http://www.nytimes.com/2010/05/21/science/21cell.html. Accessed September 1, 2016.
21. J. Craig Venter Institute. First self-replicating synthetic bacterial cell [press release] .May 20, 2010. http://www.jcvi.org/cms/press/press-releases/full-text/article/first-self-replicating-synthetic-bacterial-cell-constructed-by-j-craig-venter-institute-researcher/home/. Accessed September 1, 2016.
22. Pollack A. His corporate strategy: the scientific method. *New York Times* September 5, 2010:BU1. http://www.nytimes.com/2010/09/05/business/05venter.html?pagewanted=all. Accessed September 1, 2016.
23. TED Talk. J. Craig Venter: watch me unveil "synthetic life." http://www.ted.com/talks/craig_venter_unveils_synthetic_life.html. Accessed September 1, 2016.
24. Hutchison CA, Peterson SN, Gill SR, et al. Global transposon mutagenesis and a minimal Mycoplasma genome. *Science.* 1999;286(5447):2165-2169.
25. J. Craig Venter Institute. First self-replicating synthetic

bacterial cell. Frequently asked questions. 2010. http://www.jcvi.org/cms/research/projects/first-self-replicating-synthetic-bacterial-cell/faq. Accessed October 4, 2013.

26. Wade N. Researchers say they created a 'synthetic cell.' *New York Times* May 21, 2010:A17. http://www.nytimes.com/2010/05/21/science/21cell.html?_r=0. Accessed September 1, 2016.
27. Venter JC, Gibson D. How we created the first synthetic cell. *Wall Street Journal* May 26, 2010. http://online.wsj.com/article/SB10001424052748704026204575266460432676840.html. Accessed September 1, 2016.
28. Gibson DG, Glass JI, Lartigue C, et al. Creation of a bacterial cell controlled by a chemically synthesized genome. *Science* 2010;329(5987):52-56.
29. Macrae F. Scientist accused of playing God after creating artificial life by making designer microbe from scratch – but could it wipe out humanity? *Daily Mail* June 3, 2010. http://www.dailymail.co.uk/sciencetech/article-1279988/Artificial-life-created-Craig-Venter–wipe-humanity.html#ixzz47LQtFI7s. Accessed September 1, 2016.
30. Ibrahim MB. Craig Venter: playing God? *Muslim Matters* October 27, 2007. http://muslimmatters.org/2007/10/27/craig-venter-playing-god/. Accessed September 1, 2016.
31. President Barack Obama. Letter from President Obama to Dr. Amy Gutmann, Chair of the Presidential Commission for the Study of Bioethical Issues. May 20, 2010. http://bioethics.gov/sites/default/files/news/Letter-from-President-Obama-05.20.10.pdf. Accessed September 1, 2016.
32. Hotz RL. Scientists create synthetic organism. *Wall Street Journal* May 21, 2010. http://online.wsj.com/article/

SB10001424052748703559004575256470152341984.html. Accessed September 1, 2016.
33. Bedau M, Church G, Rasmussen S, et al. Life after the synthetic cell. *Nature* 2010;465(7297):422-424.
34. Calvert J, Martin P. The role of social scientists in synthetic biology. Science & Society Series on Convergence Research. *EMBO Rep* 2009;10(3):201-204.
35. Church GM, Regis E. *Regenesis: How Synthetic Biology Will Reinvent Nature and Ourselves*. New York: Basic Books; 2012.
36. Cho MK, Magnus D, Caplan AL, McGee D. Policy forum: genetics. Ethical considerations in synthesizing a minimal genome. *Science* 1999;286(5447):2087, 2089-2090.
37. Kruglinski S, Long M. The 10 most influential people in science. *Discover Magazine* December 2008. http://discovermagazine.com/2008/dec/26-the-10-most-influential-people-in-science. Accessed September 1, 2016.
38. Rabinow P, Bennett G. Synthetic biology: ethical ramifications 2009. *Syst Synth Biol* 2009;3(1-4):99-108.
39. Rabinow P, Bennett G. *Designing Human Practices: An Experiment with Synthetic Biology.* Chicago: University of Chicago Press; 2012.
40. Shapira P, Gök A. UK synthetic biology centres tasked with addressing public concerns. *Guardian* January 30, 2015. http://www.theguardian.com/science/political-science/2015/jan/30/uk-synthetic-biology-centres-tasked-with-addressing-public-concerns. Accessed September 1, 2016.
41. Alfred P. Sloan Foundation. Synthetic biology. http://www.sloan.org/major-program-areas/recently-completed-programs/synthetic-biology/. Accessed April 29, 2016.
42. Lempinen EW. FBI, AAAS collaborate on ambitious outreach to biotech researchers and DIY biologists. April 1, 2011. http://www.aaas.org/news/fbi-aaas-collaborate-ambitious-outreach-biotech-researchers-and-diy-biologists. Accessed September 1, 2016.

43. Parens E, Johnston J, Moses J. Ethics. Do we need "synthetic bioethics"? *Science* 2008;321(5895):1449.
44. Erik Parens, Josephine Johnston, Moses J. *Ethical Issues in Synthetic Biology: An Overview of the Debates.* Woodrow Wilson International Center for Scholars; June 2009. http://www.synbioproject.org/site/assets/files/1335/hastings.pdf. Accessed September 1, 2016.
45. Kaebnick G, Murray TH, Parens E. Ethical issues in synthetic biology. The Hastings Center; 2009. http://www.thehastingscenter.org/Research/Archive.aspx?id=1548. Accessed September 1, 2016.
46. Synthetic Biology Project. http://www.synbioproject.org/about/. Accessed August 9, 2016.
47. Synthetic Biology Project. *The Synthetic Biology Project at the Wilson Center: Eight Years of Engagement and Analysis.* Woodrow Wilson International Center for Scholars; February 2016. http://www.synbioproject.org/publications/eight-years-of-engagement-and-analysis/. Accessed September 1, 2016.
48. The National Academies. Forum on Synthetic Biology. http://sites.nationalacademies.org/pga/stl/synbio_forum/. Accessed April 30, 2016.
49. Joyce S, Mazza A-M, Kendall S; Committee on Science, Technology, and Law; Policy and Global Affairs; Board on Life Sciences, Division on Earth and Life Studies; National Academy of Engineering; National Research Council. *Positioning Synthetic Biology to Meet the Challenges of the 21st Century: Summary Report of a Six Academies Symposium Series.* Washington, DC: National Academies Press; 2013.
50. Choffnes ER, Relman DA, Pray LA; Forum on Microbial Threats; Board on Global Health; Institute of Medicine. *The Science and Applications of Synthetic and Systems Biology: Workshop Summary*. Washington, DC: National Academies Press; 2011.
51. Committee on a New Government-University Partnership

for Science and Security; Committee on Science, Technology, and Law; Policy and Global Affairs; National Research Council. *Science and Security in a Post 9/11 World: A Report Based on Regional Discussions Between the Science and Security Communities*. Washington, DC: National Academies Press; 2007.

52. Garfinkel MS, Endy D, Epstein GL, Friedman RM. *Synthetic Genomics: Options for Governance.* JCVI, CSIS, MIT; October 7, 2007. http://www.jcvi.org/cms/fileadmin/site/research/projects/synthetic-genomics-report/synthetic-genomics-report.pdf. Accessed September 1, 2016.
53. Department of Health and Human Services. Screening framework guidance for providers of synthetic double-stranded DNA. *Federal Register* 2010;75(197):62820-62832. http://www.gpo.gov/fdsys/pkg/FR-2010-10-13/html/2010-25728.htm. Accessed September 1, 2016.
54. International Gene Synthesis Consortium (ICSC). Harmonized Screening Protocol: Gene Sequence & Customer Screening to Promote Biosecurity. 2009. http://www.genesynthesisconsortium.org/Harmonized_Screening_Protocol_files/IGSC%20Harmonized%20Screening%20Protocol.pdf. Accessed September 1, 2016.
55. J. Craig Venter Institute. Policy center. http://jcvi.org/cms/research/groups/policy-center/. Accessed September 1, 2016.
56. Presidential Commission for the Study of Bioethical Issues. *New Directions: The Ethics of Synthetic Biology and Emerging Technologies.* Washington, DC: Presidential Commission for the Study of Bioethical Issues; 2010. http://bioethics.gov/sites/default/files/PCSBI-Synthetic-Biology-Report-12.16.10.pdf. Accessed September 1, 2016.
57. Gutmann A. The ethics of synthetic biology: guiding principles for emerging technologies. *Hastings Cent Rep* 2011;41(4):17-22.

58. Presidential Commission for the Study of Bioethical Issues. President's bioethics commission releases report on synthetic biology [press release]. December 16, 2010. http://bioethics.gov/node/750. Accessed September 1, 2016.
59. Pollack A. Presidential bioethics panel gives a green light to research in synthetic biology. *New York Times* December 16, 2010:A28. http://www.nytimes.com/2010/12/17/health/17synthetic.html. Accessed September 1, 2016.
60. Letter to Dr. Amy Gutmann, Chair, Presidential Commission for the Study of Bioethical Issues. December 16, 2010. http://www.foe.org/sites/default/files/Letter_to_Commission_Synthetic_Biology.pdf. Accessed September 1, 2016.
61. ETC Group. *Extreme Genetic Engineering: An Introduction to Synthetic Biology.* January 2007. http://www.etcgroup.org/content/extreme-genetic-engineering-introduction-synthetic-biology. Accessed September 1, 2016.
62. Thomas J. Beware bubbles and echo chambers. *Hastings Cent Rep* 2014;44(6 Spec no.):S43-S45.
63. Berg P, et al. Potential biohazards of recombinant DNA molecules. *Science* 1974;185(4148):303.
64. Berg P, Singer M. The recombinant DNA controversy: twenty years later. *Biotechnology (N Y)* 1995;13(10):1132-1134.
65. Bennett G, Gilman N, Stavrianakis A, Rabinow P. From synthetic biology to biohacking: are we prepared? *Nat Biotechnol* 2009;27(12):1109-1111.
66. Jasanoff S. Technologies of humility: citizen participation in governing science. *Minerva* 2003;41(3):223-244.
67. Schoch-Spana M. Public engagement and the governance of gain-of-function research. *Health Secur* 2015;13(2):69-73.
68. Grogan CM. Public engagement and the importance of content, purpose, and timing. *Hastings Cent Rep* 2014;44:S40-S42.

69. Dietz T, Stern PC, eds; Panel on Public Participation in Environmental Assessment and Decision Making; National Research Council. *Public Participation in Environmental Assessment and Decision Making*. Washington, DC: National Academies Press; 2008.
70. The Hastings Center. Hastings Center to address profound questions about human gene editing. 2016. http://www.thehastingscenter.org/for-media/press-releases/4-14-16-hastings-to-address-profound-questions-about-human-gene-editing/. Accessed September 1, 2016.
71. LEAP: Synthetic Biology Leadership Excellence Accelerator Program. 2016. http://synbioleap.org/about/. Accessed August 8, 2016.

CHAPTER 5.

ON US LEADERSHIP AND COMPETITIVENESS

The promise of the synthetic biology field is immense. According to the US National Bioeconomy Blueprint (2012), synthetic biology and related biotechnologies "can allow Americans to live longer, healthier lives, reduce our dependence on oil, address key environmental challenges, transform manufacturing processes, and increase the productivity and scope of the agricultural sector while growing new jobs and industries."[1(p1)] Specific applications are diverse and could have far-reaching effects on medicine, manufacturing, fuel production, and future biological research and exploration, including making a synthetic version of an antimalarial compound that is difficult to harvest from nature; synthesis of influenza vaccines that could be produced in a shorter time than by traditional methods; manufacturing standard biological parts that can be assembled into genetic machines; or the development of cell factories to produce biofuels and other compounds.[2,3] Synthetic biology has also brought about new tools to manipulate the genomes of biological organisms used by scientists in disparate fields and in a variety of biotechnology companies. Thanks to advances in computing power; the ability to rapidly, reliably, and inexpensively synthesize long tracts of DNA; new tools to reliably edit genomes; an increased understanding of biological systems; and the enthusiasm of young scientists who want to enter the field, synthetic biology is poised to revolutionize medicine and manufacturing.[4-7]

In spite of such promise, there are risks to US national security. These risks have generated a great deal of attention from national security experts and the US government since the inception of the field. Previous chapters have described the risks of deliberate, accidental, and thoughtless misuse of these technologies and what can be done to mitigate those risks.

But there is another scenario that would have serious negative consequences for US national security that should be considered by US policymakers and experts and should inspire action: that the United States may lose its competitive edge in synthetic biology and related technologies. While the synthetic biology field was pioneered in the United States, and the United States is currently the leader in these technologies, other nations are investing heavily in these technologies in hopes of capitalizing on the field's progress, boosting their economies, and leading the field. Some, like China, India, and the UK, have even developed specific synthetic biology roadmaps for development.[8] At the same time as there is heavy investment in synthetic biology by other nations, there is mounting concern that the competitive position of US life sciences is diminishing.[9,10]

If the United States were to lose its competitive edge in synthetic biology and related technologies, there would be serious consequences for national security. Some negative effects would be strictly economic, resulting in a declining environment for businesses and workers to be productive in synthetic biology–related industries in the long term.[11] This is important for national security because, as described in the US National Security Strategy (2015), "In addition to being a key measure of power and influence in its own right, [a strong economy] underwrites our military strength and diplomatic influence. A strong economy, combined with a prominent US presence in the global financial system, creates opportunities to advance our security."[12(p15)] Current forecasting would suggest that a loss of economic opportunities in synthetic biology could be enormous: Fidelity Investments describes synthetic biology as "the defining technology of next century" for global investments.[13] In 2012, the World Economic Forum ranked synthetic biology as the second key technology for the 21st century, after informatics.[14] According to BCC research, a market analysis company, the synthetic biology market reached nearly $2.1 billion in 2012 and $2.7 billion in 2013. They expect the market to grow to $11.8 billion in 2018 with a compound annual growth rate of 34.4% over a 5-year period from 2013 to 2018.[15]

Losing competitiveness in synthetic biology could also limit specific security applications on the horizon that are essential for national defense. These include the development of medical countermeasures for responding to biological, chemical, or radiological weapons threats and new approaches to diagnostics. A US Department of Defense (DoD) report described how synthetic biology could bring major advances to the development of high-performance sensors, sensors for unusual signatures, clandestine sensing, and high-performance materials for national defense; these applications would not likely be available to DoD based on private sector funding alone.[16] Synthetic biology may also offer the possibility for distributed manufacturing so that critical supply chains are less vulnerable to disruptions.

Synthetic Biology, Governance, and US Participation

These next several years will likely be formative in setting the "rules of the road" for emerging synthetic biology research. Yet, the United States may be disadvantaged and limited in its ability to actively participate in fundamental conversations about the governance of synthetic biology if US experts are not technological leaders in synthetic biology, as the shaping of synthetic biology governance will be dominated by the nations and their experts who are at the leading edge of technology development. This is because formal regulations or standards usually lag well behind the development of new technologies. For a new technical area, regulations are often preceded by the development of standard practices in a field, as well as cultural expectations and safety measures. These expectations and agreements build on previous sets of regulations but take new technical possibilities and dangers into account. The rules are often created by those who are most intimately familiar with the technologies—often, the scientists who are performing the work at the leading edge of development.

In the biological sciences, the most well-known example of scientists calling attention to nascent dangers in their field and setting the standards for scientific practice occurred when the field of recombinant DNA biology, an early precursor to synthetic biology, was new. In a letter published in *Science* in 1974, leading scientists and Nobel laureates recommended that certain types of recombinant DNA experiments—those with toxins, oncogenic viruses, and antibiotic resistance—should be off limits until their safety could be evaluated and assessed in a conference held a year later.[17] That conference, held at Asilomar, California, in February 1975 and attended by scientists, government officials, and members of the press, led to a lifting of the moratorium in 1976, as well as the creation of a new regulatory system for recombinant DNA work funded by the US government.[17] Efforts of the scientists to self-govern may well have forestalled restrictive national legislation.[18] Asilomar now symbolizes scientists' attention to the public's concerns, as well as the scientific community's capacity to self-govern.

A more recent example of self-governance can be found in a synthetic biology application: commercial DNA synthesis.

Companies that sell DNA synthesis products now screen their orders to determine whether a customer is ordering genetic material for dangerous pathogens and to block orders if the customer is not authorized. This screening system was developed in large part through self-governance of the commercial suppliers and interested scientists, with funding from the Alfred P. Sloan Foundation, and was eventually put into formal guidance from the US Department of Health and Human Services in 2010.[19,20]

In the synthetic biology field, there are other applications at the leading edge of development that will require governance measures to be safely and ethically applied, and some scientists have already stepped in to propose self-governance measures to deal with them. One example is the development of gene drives, which are systems that can spread a particular gene throughout a population with non-Mendelian inheritance—that is, much faster than would occur naturally.[21] These have become much easier to construct using a new gene-editing technique—clustered regularly interspaced short palindromic repeats (CRISPR/Cas9 or Cpf1)—which allows sections of DNA to be searched for and replaced in a manner roughly analogous to editing a document in Word. Some scientists have proposed using gene drives to change the DNA of mosquitoes to make them resistant to malaria. Such a project could decrease the prevalence of malaria, which currently kills more than 600,000 people—mostly children—per year. Or it could be used to target Zika. Yet, this technology could be misapplied or result in a consequential accident should the genes spread to other species or cause other unintended effects. Those scientists who have been leading the development of gene drive and gene editing technologies have also taken the lead in thinking about the safety consequences, and they have been developing a series of commonly agreed upon safeguards for laboratory research into gene drives, such as using a combination of multiple stringent confinement strategies, as any single confinement strategy could fail.[21] Scientists have also put forward ideas for how to safely use them outside of the laboratory.[22] In guidance published in June 2016, the National Academy of Science recommended a phased approach to research, testing, and evaluation in order to explore the potential benefits and reduce risks to humans and ecology.[23]

Another contentious application of synthetic biology that will require careful planning and safety standards is human germline editing, wherein modifications to sperm or egg DNA would not be applied to just 1 person, but to all their progeny. A group of interested and involved scientists met in Napa, California, to consider the ethical and safety ramifications of this work; the meeting was convened by Jennifer Doudna, one of the molecular biologists credited with developing the CRISPR/Cas9 tool. The meeting was intended to discuss the "scientific, medical, legal, and ethical implications of these new prospects for genome biology," and they identified steps so that this technology could be performed "safely and ethically."[24(p36)] In their consensus paper, published in *Science*, they recommend that the practice of germ-line editing be strongly discouraged for now, that forums be held in which this application can be discussed more broadly, and that foundational research that does not cross the line into embryo modification be encouraged.[24] The National Academy of Science also launched an initiative to recommend guidelines for the new genetic technology, to explore the scientific, ethical, and policy issues associated with human gene-editing research.[25]

Determining what the "red line" is for allowable, critical, or ethical applications of synthetic biology, as well as how much safety data are required before pressing ahead, will always be a challenging exercise, and not all scientists, experts, and observers will agree. Tension over what is acceptable to pursue has already come up for germline editing, after a Chinese research group reported that they used CRISPR techniques to modify human embryos in an attempt to correct the mutation that leads to beta thalassemia, a blood disorder.[26] A more recent study from a different research group in China attempted to modify human embryos to make them resistant to infection with HIV.[27](And there are at least 3 additional research groups in China known to be pursuing gene editing in human embryos.) While the standards or expectations set by the scientific community will be impossible to enforce in an international context, the scientific community does set boundaries; those who flout those standards have to justify their actions in the international practice of science, and those boundaries and expectations are set by the leaders in the field. In the case of germline editing, the Chinese research on beta thalassemia was rejected by top-tier scientific journals *Nature* and *Science,* in part because of ethical objections.[28]

Self-governance of science has its critics, who are justifiably skeptical that scientists can be trusted to govern their own research fairly and who question the effectiveness of this approach in an international context, as the embryo editing example illustrates. However, self-governance is not the sole mechanism of governance in this area, as many foundational aspects of biotechnology and laboratory practice are already tightly regulated, and also because in forming new rules there is often a complex interplay among scientists, journalists, and policymakers to bring about new guidelines. In the case of DNA synthesis guidance, while there was substantial work done by scientists and interested parties to prevent misuse of DNA synthesis and promote screening, the issue became more salient, requiring immediate action, after a journalist ordered a small segment of DNA that encoded the smallpox virus.[29] Still, feasible alternatives to self-governance are limited when technologies are still in the early stages of development, particularly when the applications are of broad interest, generating funding from private companies and multiple national governments, when the work is pursued in many places internationally, and when the technologies have great potential for tangible benefits to health and medicine. In addition, the amount of technical knowledge required for understanding the implications of new research and what can be done to ameliorate negative consequences makes it challenging even for scientists in distinct disciplines to evaluate research outside their expertise, because understanding the technical details inherent in the technology are critical both for identifying problems as well as proposing solutions.

There are additional applications of synthetic biology that have already generated conversations about governance in the scientific community—such as rescuing a species on the path to extinction; or even using synthetic biology for "de-extinction," to bring back a species that was lost because of human hunting or negligence; or brewing opiates by fermentation in a process not unlike brewing beer.[30-32] These applications have already sparked scientific involvement in discussions of what is technically possible and what rules should be developed. In 5 to 10 years, the list of applications that will require expert opinion and involvement to set expectations, standards of practice, and self-governance may well be very different, just as consequential, and require technical experts to take the lead in setting norms and safety standards. If US scientists, policymakers, and institutions would like to have some say in what is decided, they will need to be at the forefront of those technologies.

US S&T: Evidence of Promise and Decline

The United States is currently a leader in synthetic biology, as well as biotechnology and biomedical research, and it is the focus of a great deal of private sector investment; these investments may help to bring at least 100 products to the market in the near future.[33,34] According to a DoD report, the US government also provides at least $220 million annually toward synthetic biology R&D, with investments from the Department of Energy, the National Science Foundation (NSF), the DoD (including DARPA), the National Institutes of Health (NIH), and the US Department of Agriculture (USDA).[16] An analysis from the Wilson Center found that between 2008 and 2014, the US government invested a total of $820 million in synthetic biology research, with DARPA funding nearly $110 million in 2014.[34] Indeed, synthetic biology researchers in the United States have largely relied on DARPA funding, such as in their Living Foundries program, which aims "to create a revolutionary, biologically-based manufacturing platform to provide access to new materials, capabilities and manufacturing paradigms for the DoD and the Nation.[35]

The United States does not have a specific synthetic biology technology roadmap, but on April 27, 2012, the Obama administration released their National Bioeconomy Blueprint, "a comprehensive approach to harnessing innovations in biological research to address national challenges in health, food, energy, and the environment."[36] The blueprint identifies the administration's priorities to grow the bioeconomy through increased investment in research and development, expansion of public-private partnerships, and regulatory reform and, in numerous instances, specifically mentions the enormous promise of synthetic biology. While the government programs and initiatives listed in the Bioeconomy Blueprint were already in progress, the blueprint served as a sign of federal commitment to developing the biological sciences as a component of the US economy.[37]

The United States also has a robust bioeconomy, which includes synthetic biology and related technologies. Defining the economic impact of synthetic biology is difficult, as "traditional" biotechnologies are also taking advantage of pervasive synthetic biology techniques, and the components of the biological economy are not as accurately tracked as other sectors of the

US economy.[34] Looking at the bioeconomy as a whole, Robert Carlson, an industry analyst, found that products derived from biology contributed an estimated $350 billion to American GDP in 2012, and the "bioeconomy" grew 15% annually and accounted for nearly 7% of total US GDP growth in 2011 and 2012.38 Engineered organisms led to products worth more than $350 billion per year to the US economy. DuPont, Pfizer, Bausch & Lomb, Coca-Cola, and other Fortune 500 companies either make or use products derived from engineered organisms, including food, clothing, medicines, and beauty products.[16] For example, DuPont has been producing commercial quantities of the polymer 1,3-propanediol from engineered bacteria since 2006, which is 37% of the material in their Sorona fibers—used for everything from carpets to car interiors.[39] Some investors forecast the possibility of billions of dollars of growth in American manufacturing through the biotechnology sector, including at the Goodyear Tire & Rubber Company, DuPont, Archer Daniels Midland, and Solazyme.[39]

Yet, in spite of clear US leadership in synthetic biology, there are well-documented concerns about the United States falling behind in biotechnology and in science more generally, as well as concerns about falling US biomedical research budgets, STEM (science, technology, engineering, and mathematics) workforce decline, and outsourcing by international pharmaceutical and biotechnology companies, which are applicable to synthetic biology as well. Global indicators for the biosciences and biotechnology, including R&D outputs as well as shares of the global pharmaceutical industry, higher education, and workforce, are showing what NIH called an "erosion of the competitive position of the U.S. life sciences industry over the past decade."[9(p1)] China will overtake the United States in R&D spending by 2020.[9] In 2007, China overtook the United States in the number of doctoral degrees awarded in the natural sciences and engineering.[40] Europe is thought to be the fastest growing market for synthetic biology products, and the UK is considered to be one of the most innovative and dynamic, and healthcare industries there are expected to grow in the future.[11]

US students in synthetic biology have been affected as well, as seen in the international Genetically Engineered Machine (iGEM) competition. This competition pits teams of synthetic biologists (primarily undergraduates) from all over the world in competition to engineer biological systems and operate them in living cells. It began as a small class at MIT in Cambridge, Massachusetts, in 2003 and has grown to more than 2,000 international participants and more than 16,000 alumni.[41] In 8 of the past 10 years, US student teams have failed to win "in part because of a lack of laboratory facilities" and other support.[10(p29)]

In a DoD report from the Office of Technical Intelligence, Office of the Assistant Secretary of Defense for Research and Engineering, dwindling human capital was identified as an obstacle to DoD operating effectively and efficiently in the future: "There are few highly-experienced program managers in the Department, few leading scientists, and even fewer individuals in uniform with deep knowledge of the [synthetic biology] field. The lack of uniformed expertise is particularly troubling."[16(p20)]

In contrast to other industries that require substantial natural resources, such as arable land, oil, or natural gas, synthetic biology and related technologies have few barriers to entrance, and emerging markets can become competitive quickly. Major gains have been made rapidly in several nations by changing policies and investments. Though there are several countries making substantial strategic investments in synthetic biology, the example of China is notable. The Chinese Academy of Sciences includes synthetic biology in its *Innovation 2050: Technology Revolution and the Future of China Roadmap.*[42] An example of China's substantial investments in synthetic biology is its support of the Beijing Genomics Institute (BGI), a company located in the city of Shenzhen. It is the world's largest genetic research center, with more sequencing capacity than the entire United States and about one-quarter of the total global capacity.[43,44] In 2013, BGI purchased the Mountain View, California–based company Complete Genomics, 1 of the 2 leading companies in the world that make equipment for sequencing DNA, further increasing BGI's dominance in the sequencing market. Previously known solely for their speed and proficiency in sequencing genomes, the company is starting to diversify and innovate, making several commercial diagnostic tests. The comprehensive database of sequencing information they have developed—they have sequenced many hundreds of different types of bacteria; crops such as rice, soybeans, and cucumbers; and dozens of animals including the giant panda; as well as human genomes—is seen as a springboard for new discoveries, as well as the development of new drugs and therapies. BGI has also been helpful in international science efforts, playing a role in the Human Genome Project and identifying the foodborne *Escherichia coli* outbreak in Germany that infected nearly 4,000 people, killing 53.[45,46]

China's research system still draws attention for its ethics problems, including fraudulent results, plagiarism, junk patents, and unsafe or ineffective medical practices. However, scientists who collaborate with Chinese researchers believe that the research system is changing and becoming more internationally competitive.[47,48] This change is due in part to China's successful efforts to lure back Chinese researchers who were trained and/or employed in the United States, offering them bigger budgets and greater research freedom than they would have in the United States. In the case of BGI, international collaborations are integral to their success and include partnering with the Gates

Foundation as well as hospitals and universities in the United States and Europe.[49]

The UK has also looked to synthetic biology for economic growth and other benefits. A roadmap for synthetic biology was released in 2012, and to date the UK government has invested approximately £200 million for research and the creation of several synthetic biology research groups across the country.[50,51] In a 2012 study that mapped the scientific landscape for synthetic biology, the UK was second only to the United States in having its scientists author publications on synthetic biology.[52] The UK is also taking steps to dissociate synthetic biology from the controversies surrounding genetically modified organisms (GMOs). At the world conference on synthetic biology held at Imperial College, London, in 2013, a minister from the House of Commons told the assembled scientists, referring to GMOs, that the UK would not become "a museum of twentieth century technologies in the twenty-first century."[53] GMO restrictions have been a competitive hindrance in UK participation in the field of synthetic biology and in biotechnology in the UK and EU more generally. The effects of the UK vote to leave the European Union ("Brexit") on synthetic biology and biotechnology development are unclear but are unlikely to diminish enthusiasm for synthetic biology.[54]

Steps to Greater US Competitiveness

Measures aimed at boosting competitiveness in science and technology generally are broadly applicable for synthetic biology and should be pursued by the US government. These initiatives include increased basic research funding with minimal fluctuations from year to year, workforce development, and STEM (science, technology, engineering and mathematics) education initiatives, as well as financial incentives to start and fund biotechnology and synthetic biology companies and discourage them from locating offshore.[9-11,43,55] Some economists have recommended that foreign students who receive their PhDs for research in technical STEM-related fields at US universities should be encouraged to stay in the country to pursue their careers and receive automatic green cards enabling them to work in the United States.[11]

But to remain competitive in synthetic biology, the US government will also need to take specific action on fundamental policy issues that will affect the field's development. One priority should be responding to and countering anti-GMO sentiments and legislation, which are on the rise. The ability to specifically modify, recode, transform, and manipulate the genetic code of organisms—and thus, the characteristics of the organisms themselves—is much more powerful using synthetic biology techniques than was ever before possible. In fact, synthetic biology has been described as "genetic engineering on steroids."[56] It should thus be no surprise that long-standing debates, concerns, and activism surrounding the topic of GMOs would arise in response to synthetic biology. While the anti-GMO movement has been typically thought of as a European concern, which has diminished European agricultural competitiveness and has thus given the United States a competitive edge, there are warning signs that anti-GMO concerns are growing and will no longer be possible for scientists and policymakers in the United States to ignore. Simply put, concerns about GMOs that cannot be scientifically justified are at odds with US competitiveness in synthetic biology and other biotechnologies. The United States should actively counter anti-GMO policies, while also ensuring that synthetic biology is appropriately regulated, and work to inform the public about how products are regulated for safety.

The United States' approach to the regulation of biotechnology, different from that of Europe, has so far carried over to the regulation of synthetic biology applications. The focus of regulation and safety in the United States has traditionally been on the end result: the *product*. This is not to say that all conceivable GMO products are guaranteed to be safe, but it is the product that should be subject to a safety determination, not the process used to make it, whether that process is synthetic biology or another technique.

In contrast to the United States, European regulatory agencies have typically embraced the "precautionary principle," which places the burden of proof on the developer of a product to show that the process used to make a particular product is not harmful. There are multiple formulations of the precautionary principle; one often-used definition came from the Wingspread Conference on the Precautionary Principle in 1998 and states:

"When an activity raises threats of harm to human health or the environment, precautionary measures should be taken even if some cause and effect relationships are not fully established scientifically. . . . The process of applying the Precautionary Principle must be open, informed and democratic and must include potentially affected parties. It must also involve an examination of the full range of alternatives, including no action."[57]

While precaution in the face of indeterminate risks sounds to many a reasonable approach—it is the essence of the expression "look before you leap"—in practice, critics have charged that the usual result of its application is inaction.[58,59] In the case of synthetic biology, a precautionary approach would result in a general moratorium on the release and commercial use of synthetic biology until there is a research agenda, alternative approaches have been fully considered, a technology assessment has been performed, and there is national and perhaps international oversight for each of the technologies.[60] This could take many years even if all nations were in agreement about the need for it, which they are not.

The distrust of GMOs has had a detrimental economic effect in the EU. The prohibition on GMOs in the EU decreases profit margins for European farmers by up to a billion dollars each year.[61] The British government Biotechnology and Biological Sciences Research Council (BBRC) has charged that the precautionary approach has "effectively stifled GM crop farming in the EU."[62] It costs £10-20 million more to put a GM crop through an EU approval process than for conventionally bred new crops.[63] A group of 21 prominent plant scientists wrote an open letter stating that Europe will lose research prominence unless field trials are allowed of GM crops and that they will fall short of producing "world-class science" unless a pro-science stance is taken by policymakers.[64] Science advisors to former British Prime Minister David Cameron called for scrapping "dysfunctional EU regulations" around GMOs, and they note the hypocrisy in that the EU imports 70% of its animal feed, most of it made with GMOs. The United States, Canada, Brazil, and Argentina grow 90% of the planet's GM crops.[65]

It should be stated that the evidence on the safety of "GMO" foods is in, and the results are clear. Genetic engineering presents no unique hazards compared to other methods that create genetic modification, such as traditional breeding or hybridization. Major scientific organizations, including the American Association for the Advancement of Science (AAAS), the National Academies of Science, and the American Medical Association (AMA) all back GMOs as being safe. In a meta-review of the safety of genetically engineered crop research that evaluated 1,783 research papers and reports from the years 2002 to 2012, no significant hazards were identified.[66,67] The European Commission funded 1,340 research projects from 500 independent teams looking at GMO safety and none found risks.65 In addition to the lack of harm found in GMO use, there are substantial benefits to using GMOs: lower food prices; less pesticide use, which is safer for farmers; less water needed;

increased crop yields; and more stable prices.[65] There is also necessity: The UN FAO estimates that the world will need to grow 70% more food by 2050 just to keep up with population growth. There may be 10 billion people on earth, requiring more food to be grown in the next 75 years than has been produced in all of human history.[68] Climate change, with the loss of arable land, will worsen this problem. Maximizing food production through GMOs may be the only avenue to provide people with enough food.

The anti-GMO movement has also cost lives. Vitamin A deficiencies cause more than 1 million deaths every year, as well as half a million cases of irreversible blindness.[65] In spite of this, the GMO Golden Rice, engineered to deliver more vitamin A than spinach, has not been allowed to be grown in India and the Philippines, largely due to the activities of Greenpeace and other anti-GMO organizations.[69] Kenya had an outright ban on GMOs in spite of an advancing crop disease that affects corn, the maize lethal necrosis disease, which could lead to food insecurity and famine as crops are destroyed by the virus.[70] Kenyan officials now say the ban resulted from their being misled by French activists who claimed that GM products cause tumors and were unfit for human consumption; the ban on GMOs was expected to be lifted by the end of 2015 but remains in place as of this writing.[71,72]

There is cause for concern that anti-GMO sentiments are increasing in the United States and will harm US competitiveness, particularly when it comes to realizing beneficial synthetic biology applications. In the United States, the use of anti-GMO sentiment as a marketing tool has been growing. Products that are marketed as not containing GMOs will account for 30% of US food and beverage sales by 2017.[73,74] Whole Foods started labeling their products that are GMO-free, stating that they were responding to their customers, "who have consistently asked us for GMO labeling and we are doing so by focusing on where we have control: in our own stores."[75] By 2018, all products in their US and Canadian stores will be labeled to indicate if they contain GMOs. This is the first national grocery chain to set a deadline for "full GMO transparency."[75] Chipotle and Trader Joe's also have decided to not sell foods made with GMOs and to use this fact in advertising campaigns.

Congress established the National Organic Standards Board (NOSB) under the USDA through the Organic Food Production Act, and it was charged with developing standards, which have become known as the "Organic Rule." The Organic Rule expressly forbids the use of GMO crops, antibiotics, and synthetic nitrogen fertilizers, as well as food additives and ionizing radiation. The Organic Seal is a marketing tool and is separate from safety. But organic marketers represent conventionally grown or GM crops as dangerous.[76] Major scientific organizations have tended to be against labeling laws because of what happened in Europe: In 1997, when there was growing opposition to GMOs in Europe, the EU began to require labels. By 1999, to avoid the GMO labels, most European retailers had removed those ingredients, and now GM products cannot be found in European stores.[69]

At least 20 states were considering GM labeling bills; most people in favor of labeling would use those labels to avoid eating those foods.[65,68] Connecticut, Maine, and Vermont have passed labeling laws. In fact, the Vermont law was the catalyst for a federal bill expected to go into law, which will render the state laws null and void.[77] The federal law requires food to be labeled if designated as a GMO product, but the measure is significantly less burdensome for food and biotechnology companies than the Vermont law. It is expected, however, that the labeling fight will continue in the regulatory realm, through decisions that will need to be made by the FDA and USDA to implement the federal law.[78]

Anti-GMO groups have already found synthetic biology as a target. One example comes from Ecover, a Belgian company that makes detergents, and Method, which is a subsidiary company. Ecover purchased oils for its products developed by Solazyme, a US company that uses synthetic biology to produce an environmentally sustainable substitute for palm kernel oil in algae. Palm kernel oil is in high demand, which has led to conservationist concerns about overcultivation, deforestation, and loss of tropical habitats. Ecover found itself inundated with petitions to stop using synthetic biology for using what an anti-GMO group labeled an "extreme biotech oil."[79] Synthetic vanillin, produced by Evolva, a Swiss synthetic biology company that partnered with International Flavors & Fragrances (IFF-USA), has met with a similar response from anti-GMO groups.[80] Friends of the Earth (FOE) "persuaded" Haagen-Dazs not to use vanillin made through synthetic biology, but that was not likely to occur anyway, as Haagen-Dazs uses only vanilla extract from vanilla beans. It is another example of the cynical use of anti-GMO sentiment for marketing purposes.[80,81]

If anti-GMO sentiment increases, there will be a great deal of pressure placed on policymakers and regulators to adhere to the precautionary principle. Communicating the science behind GMOs is a much more difficult task than simply labeling it as bad, and the United States is not immune from applying a more precautionary stance to regulatory areas.[82] Still, resisting efforts to undermine a positive future for synthetic biology is critical for US competitiveness, as is making sure that synthetic biology products are, indeed, appropriately regulated. While the product, not the process, should be the focus of regulation and oversight, at this time there are gaps in regulation, and synthetic biology is likely to increase them.[83] As one example, a 2013 fundraising campaign on Kickstarter caused consternation by promising to produce glowing plants and distributing seeds to more than 8,000 supporters.[39] The mechanisms used to produce the plants, distribute them, and plant them did not violate any current rules or regulations. The technological challenges have also proven to be much higher than expected, so no plants have thus far been distributed.[84] Nonetheless, allowing glowing plants to be introduced into the environment without regulatory review struck many as a foolhardy proposal and risked bringing about negative public opinions about synthetic biology.[76] Current oversight depends on whether plant pests or some plant pest component is used for engineering the plant. As many newer methods of genetic manipulation would not involve such a step, this would leave many engineered plants without regulatory review before they are cultivated in the environment for field trials or commercial production.[83]

There is an opportunity to make the regulation more coherent. In July 2015, the White House directed the 3 federal agencies that have oversight responsibilities for biotechnology products—the Environmental Protection Agency (EPA), the FDA, and the USDA—to develop a long-term strategy for the oversight of future products in biotechnology and to update what is known as the "Coordinated Framework." The Coordinated Framework for the Regulation of Biotechnology was introduced in 1986 by the White House Office of Science and Technology Policy (OSTP) as a comprehensive federal regulatory policy to ensure the safety of biotechnology products beyond pharmaceuticals; it was last updated in 1992. Updating the framework became necessary, as it was outdated and confusing, and its complexity made it "difficult for the public to understand how the safety of biotechnology products is evaluated," as the glowing plant example makes clear.[85] In addition, the regulatory process could be unnecessarily challenging for small companies. The Coordinated Framework will be updated and will clarify which agencies have responsibility to regulate products that might fall under authorities of multiple agencies.[85] In addition to this work, there will be a long-term strategy developed with an aim of making sure that the regulatory system is well-equipped to assess the risks associated with future biotechnology products. The National Academies of Sciences, Engineering, and Medicine have also been commissioned to perform an outside, independent analysis of the future landscape of biotechnology products.[85]

Engaging in International Discussions

Formal mechanisms of international governance of synthetic biology need to be addressed by the US government. Synthetic biology has become a major topic in the Convention on Biological Diversity (CBD), which has 168 member nations but does not include the United States, which has signed but not ratified the treaty. The Cartagena Protocol in the CBD provides an international regulatory framework for the transfer, handling, and use of living modified organisms (LMOs) resulting from modern biotechnology. At the CBD 10th Conference in 2010, the members agreed that the release of products of synthetic biology requires caution and the application of the precautionary principle. Another protocol to the CBD, the Nagoya Protocol on Access to Genetic Resources and the Fair and Equitable Sharing of Benefits, aims at sharing the benefits arising from the use of genetic resources in a fair and equitable way and will also affect the synthetic biology industry.

The US government should pay great attention to the activities of this treaty, to minimize the impact of restrictions on developing synthetic biology technologies. Although the US is not bound by activities or resolutions of the convention, the synthetic biology market will be affected if the United States and the scientific community do not become more engaged in the CBD process.[86] The precautionary stance that the treaty parties are taking, as well as the possible consideration to bar some genetic sequences for use, may limit US synthetic biology exports and could hamper the field's development of beneficial applications.[87] The United States should work with other nations that are party to the Convention on Biological Diversity or the Nagoya Protocol to minimize the impact on US economic interests. At the heart of the treaty is a justifiable concern about the fair and equitable sharing of benefits arising from genetic resources. Rather than closing off a potentially broadly beneficial technology, other mechanisms should be created that directly address the need for fairness and access to benefits arising from the technologies.

Conclusion

Synthetic biology is a fast-moving field, and it has already been applied to the development of new vaccines and medical countermeasures as well as the production of biofuels, detergents, adhesives, perfumes, tires, and specialized chemicals that formerly required the use of petrochemicals. As the field continues to expand, synthetic biology may become a pervasive industrial technology. Proponents believe that synthetic biology and related technologies could be the foundation of a new manufacturing economy for the United States and could contribute to industries essential to US national security. While the field was pioneered in the United States, other nations are hoping that investments in this area will boost their economies, and so the great technical lead that the United States has enjoyed will inevitably shrink. However, it is imperative that the United States not fall behind and that it makes investments to ensure that it is positioned to enjoy the fruits of a robust bioeconomy as well as participate in the technical back and forth that will set standards and limits for governance in controversial applications of the technologies. In actively taking steps to increase its global competitiveness in synthetic biology, the United States should also address fundamental policy issues about GMOs and make sure that all products are appropriately regulated—whether they are made through traditional methods, synthetic biology, or an innovative technology yet to be developed.

References

1. The White House. *National Bioeconomy Blueprint.* April, 2012. Available at http://www.whitehouse.gov/sites/default/files/microsites/ostp/national_bioeconomy_blueprint_april_2012.pdf
2. Paddon CJ, Westfall PJ, Pitera DJ, et al. High-level semi-synthetic production of the potent antimalarial artemisinin. *Nature.* Apr 25 2013;496(7446):528-532. Available at http://www.ncbi.nlm.nih.gov/pubmed/23575629
3. Schmidt M. Introduction. In: Schmidt M, ed. *Synthetic BIology: Industrial and Environmental Applications*: Wiley-Blackwell; 2012:1-6. Available at http://www.amazon.com/Synthetic-Biology-Industrial-Environmental-Applications/dp/3527331832
4. Carlson RH. *Biology is technology : the promise, peril, and new business of engineering life.* Cambridge, Mass.: Harvard University Press; 2010 p. 168-169. Available at http://www.hup.harvard.edu/catalog.php?isbn=9780674060159
5. Committee on Industrialization of Biology, Board on Chemical Sciences andTechnology, Board on Life Sciences, Division on Earth and Life Studies, National Research Council. *Industrialization of Biology: A Roadmap to Accelerate the Advanced Manufacturing of Chemicals.* 2015. Available at http://www.nap.edu/catalog/19001/industrialization-of-biology-a-roadmap-to-accelerate-the-advanced-manufacturing
6. Gigi Kwik Gronvall, Ryan Morhard, Kunal Rambhia, Anita

Cicero, Inglesby T. *The Industrialization of Biology and Its Impact on National Security.*June 8, 2012. Available at http://www.upmc-biosecurity.org/website/resources/publications/2012/pdf/2012-06-08-industrialization_bio_natl_security.pdf

7. Nancy J. Kelley & Associates. The Promise and Challenge of Engineering Biology in the United States. *Industrial Biotechnology.* June, 2014;10(3):137-139. Available at http://online.liebertpub.com/doi/abs/10.1089/ind.2014.1516
8. Organisation for Economic Co-operation and Development (OECD). *Emerging Policy Issues in Synthetic Biology.* OECD Publishing.June 4, 2014. Available at http://www.oecd-ilibrary.org/science-and-technology/emerging-policy-issues-in-synthetic-biology_9789264208421-en
9. National Institutes of Health. *Global Competitiveness– The Importance of U.S. Leadership in Science and Innovation for the Future of our Economy and our Health.*February, 2015. Available at http://www.nih.gov/about/impact/impact_global.pdf
10. MIT Committee to Evaluate the Innovation Deficit. *THE FUTURE POSTPONED: Why Declining Investment in Basic Research Threatens a U.S. Innovation Deficit.*April, 2015. Available at http://dc.mit.edu/sites/default/files/innovation_deficit/Future%20Postponed.pdf
11. Porter M, Rivkin J. What Washington Must Do Now: An Eight-Point Plan to Restore American Competitiveness. *The Economist.* November 21, 2012. Available at http://www.hbs.edu/competitiveness/Documents/theworldin2013.pdf
12. The White House. *National Security Strategy.* 2015. Available at https://www.whitehouse.gov/sites/default/files/docs/2015_national_security_strategy.pdf
13. SPONSORED CONTENT: Synthetic Biology – Fidelity

Investments. 2016; http://www.washingtonpost.com/business/sponsored-content-synthetic-biology—fidelity-investments/2012/05/02/gIQAvIgPwT_video.html. Accessed September 12, 2016.

14. Global Agenda Council on Emerging Technologies. The top 10 emerging technologies for 2012. 2012; http://forumblog.org/2012/02/the-2012-top-10-emerging-technologies/. Accessed September 12, 2016.
15. Bergin J. *Synthetic Biology: Global Markets.* BccResearch.June, 2014. Available at http://www.bccresearch.com/market-research/biotechnology/synthetic-biology-bio066c.html
16. Office of Technical Intelligence, Office of the Assistant Secretary of Defense for Research and Engineering. *Technical Assessment: Synthetic Biology.* Department of Defense Research and Engineering.January, 2015. Available at http://defenseinnovationmarketplace.mil/resources/OTI-SyntheticBiologyTechnicalAssessment.pdf
17. Berg P, et al. Potential biohazards of recombinant DNA molecules. *Science.* Jul 26, 1974;185(4148):303. Available at http://www.ncbi.nlm.nih.gov/pubmed/11661080
18. Berg P, Singer M. The recombinant DNA controversy: twenty years later. *Biotechnology (N Y).* Oct 1995;13(10):1132-1134. Available at http://www.ncbi.nlm.nih.gov/pubmed/11644758
19. Garfinkel MS, Endy D, Epstein GL, Friedman RM. *Synthetic Genomics: Options for Governance.*October 7, 2007. Available at http://www.jcvi.org/cms/fileadmin/site/research/projects/synthetic-genomics-report/synthetic-genomics-report.pdf
20. US Department of Health and Human Services. *Screening Framework Guidance for Providers of Synthetic Double-Stranded DNA.* Federal Register Volume 75, Number 197. Pages 62820-62832.October 13, 2010. Available at

http://www.gpo.gov/fdsys/pkg/FR-2010-10-13/html/2010-25728.htm
21. Akbari OS, Bellen HJ, Bier E, et al. BIOSAFETY. Safeguarding gene drive experiments in the laboratory. *Science.* Aug 28 2015;349(6251):927-929. Available at http://www.ncbi.nlm.nih.gov/pubmed/26229113
22. Oye KA, Esvelt K, Appleton E, et al. Biotechnology. Regulating gene drives. *Science.* Aug 8, 2014;345(6197):626-628. Available at http://www.ncbi.nlm.nih.gov/pubmed/25035410
23. Committee on Gene Drive Research in Non-Human Organisms: Recommendations for Responsible Conduct, Board on Life Sciences, Division on Earth and Life Studies, National Academies of Sciences, Engineering, and Medicine. *Gene Drives on the Horizon: Advancing Science, Navigating Uncertainty, and Aligning Research with Public Values* National Academies Press. 2016. Available at http://nas-sites.org/gene-drives/
24. Baltimore D, Berg P, Botchan M, et al. Biotechnology. A prudent path forward for genomic engineering and germline gene modification. *Science.* Apr 3 2015;348(6230):36-38. Available at http://www.ncbi.nlm.nih.gov/pubmed/25791083
25. Begley S. U.S. science leaders to tackle ethics of gene-editing technology. *Reuters.* May 18, 2015. Available at http://www.reuters.com/article/2015/05/18/science-genes-nas-idUSL1N0Y910020150518
26. Liang P, Xu Y, Zhang X, et al. CRISPR/Cas9-mediated gene editing in human tripronuclear zygotes. *Protein & cell.* May 2015;6(5):363-372. Available at http://www.ncbi.nlm.nih.gov/pubmed/25894090
27. Cyranoski D, Reardon S. Embryo editing sparks epic debate. *Nature.* Apr 30, 2015;520(7549):593-594. Available at http://www.ncbi.nlm.nih.gov/pubmed/25925450
28. Cressey D, Cyranoski D. Gene editing poses challenges for

journals. *Nature.* Apr 30, 2015;520(7549):594. Available at http://www.ncbi.nlm.nih.gov/pubmed/25925451

29. Randerson J. Did anyone order smallpox? *Guardian Weekly.* 2006. Available at http://www.theguardian.com/science/2006/jun/23/weaponstechnology.guardianweekly
30. Oye KA, Lawson JC, Bubela T. Drugs: Regulate 'home-brew' opiates. *Nature.* May 21 2015;521(7552):281-283. Available at http://www.ncbi.nlm.nih.gov/pubmed/25993942
31. Church G. De-extinction is a Good Idea. *Scientific American* August 26, 2013. Available at http://www.scientificamerican.com/article.cfm?id=george-church-de-extinction-is-a-good-idea
32. Lajoie MJ, Kosuri S, Mosberg JA, Gregg CJ, Zhang D, Church GM. Probing the Limits of Genetic Recoding in Essential Genes. *Science.* October 18, 2013;342(6156):p. 361-363. Available at http://www.sciencemag.org/content/342/6156/361.short
33. OECD. *Emerging Policy Issues in Synthetic BIology.* OECD Publishing.June 4, 2014. Available at http://www.oecd-ilibrary.org/science-and-technology/emerging-policy-issues-in-synthetic-biology_9789264208421-en
34. *U.S. Trends in Synthetic Biology Research Funding.* Wilson Center September 2015. Available at http://www.synbioproject.org/site/assets/files/1386/final_web_print_sept2015.pdf?
35. Defense Advanced Research Projects Agency (DARPA). Living Foundries. 2015; http://www.darpa.mil/program/living-foundries. Accessed September 30, 2015.
36. Office of Science and Technology Policy. *Obama Administration Unveils "Bioeconomy Blueprint" Announces New R&D Investments* April 26, 2012. Available at https://www.whitehouse.gov/sites/default/files/microsites/ostp/bioeconomy_press_release_0.pdf
37. Pollack A. White House Promotes a Bioeconomy. *The New York Times.* April 26, 2012. Available at

http://www.nytimes.com/2012/04/26/business/energy-environment/white-house-promotes-a-bioeconomy.html?_r=0
38. Carlson R. The U.S. Bioeconomy in 2012 reached $350 billion in revenues, or about 2.5% of GDP. 2014; http://www.synthesis.cc/2014/01/the-us-bioeconomy-in-2012.html. Accessed September 30, 2015.
39. Evans A. Glowing Plants: Natural Lighting with no Electricity. 2013; https://www.kickstarter.com/projects/antonyevans/glowing-plants-natural-lighting-with-no-electricit. Accessed September 12, 2016.
40. National Science Foundation. *Science and Engineering Indicators 2014.* 2014. Available at http://www.nsf.gov/statistics/seind14/index.cfm/chapter-4/c4h.htm
41. Richard A. Johnson, National Academy of Sciences, Board on Life Sciences, NAS Forum on Synthetic Biology. Synthetic Biology: 10 Policy Reasons It Matters to U.S. Foreign Policy *The Evolving Nature of Synthetic Biology: A Panel Discussion on Key Science, Policy, and Societal Challenges Facing the International Community*. The State Department, Washington, DC. 2013. Available at http://sites.nationalacademies.org/cs/groups/pgasite/documents/webpage/pga_084543.pdf
42. American Association of State Colleges and Universities. *An urgent imperative: proceedings of the Wingspread Conference on Teacher Preparation*. Washington, DC: American Association of State Colleges and Universities; 1986. Available at https://searchworks.stanford.edu/view/1639638
43. Atkinson RD, Ezell SJ, Giddings LV, Stewart LA, Andes SM. *Leadership in Decline: Assessing U.S. International Competitiveness in Biomedical Research.*May, 2012. Available at http://www.unitedformedicalresearch.com/wp-content/uploads/2012/07/Leadership-in-Decline-Assessing-US-International-Competitiveness-in-Biomedical-Research.pdf

44. Specter M. The Gene Factory. *The New Yorker.* 2014. Available at http://www.newyorker.com/magazine/2014/01/06/the-gene-factory
45. Turner M. E. coli outbreak strain in genome race. *Nature News.* July 21 2011. Available at http://www.nature.com/news/2011/110721/full/news.2011.430.html
46. European Food Safety Authority. *E. coli: Rapid response in a crisis.* July 11, 2012. Available at http://www.efsa.europa.eu/en/press/news/120711
47. Cyranoski D. Research ethics: Zero tolerance. *Nature News.* January 11 2012. Available at http://www.nature.com/news/research-ethics-zero-tolerance-1.9756
48. Looks good on paper. *The Economist.* September 28, 2013. Available at http://www.economist.com/news/china/21586845-flawed-system-judging-research-leading-academic-fraud-looks-good-paper
49. Sender H. Chinese innovation: BGI's code for success. *Financial Times.* February 16, 2015. Available at http://on.ft.com/2dohvI0
50. UK synthetic biology roadmap coordination group. *A synthetic biology roadmap for the UK.* 2012. Available at http://www.rcuk.ac.uk/documents/publications/SyntheticBiologyRoadmap.pdf
51. Shapira P, Gök A. UK Synthetic Biology Centres tasked with addressing public concerns. *The Guardian.* January 30, 2015. Available at http://www.theguardian.com/science/political-science/2015/jan/30/uk-synthetic-biology-centres-tasked-with-addressing-public-concerns
52. Oldham P, Hall S, Burton G. Synthetic biology: mapping the scientific landscape. *PLoS One.* 2012;7(4):e34368. Available at http://www.ncbi.nlm.nih.gov/pubmed/22539946
53. *RT Hon David Willetts MP, House of Commons* BioBricks Foundation SB6.0: The Sixth International Meeting on Synthetic Biology, held July 9-11, 2013 at Imperial College, London, UK2013. Available at http://vimeo.com/70118672

54. Warmflash D EJ. How Brexit will impact the future of farming, GMOs and gene editing in Britain and Europe. *Genetic Literacy Project.* June 29, 2016. Available at https://www.geneticliteracyproject.org/2016/06/29/brexit-will-impact-future-farming-gmos-gene-editing-britain-europe/
55. IEEE-USA. *Maintaining U.S. Leadership in Innovation and Competitiveness.*November 22 2013. Available at https://www.ieeeusa.org/policy/positions/Innovation1113.pdf
56. Voosen P. Synthetic Biology Comes Down to Earth. *The Chronicle of Higher Education.* March 4, 2013. Available at http://chronicle.com/article/Synthetic-Biology-Comes-Down/137587/
57. *Wingspread Conference on the Precautionary Principle.*January 26 1998. Available at http://www.sehn.org/wing.html
58. Hahn RW, Sunstein CR. The Precautionary Principle as a Basis for Decision Making. *The Economist's Voice.* 2005;2(2). Available at http://ssrn.com/abstract=721122
59. Clarke S. New Technologies, Common Sense and the Paradoxical Precautionary Principle. In: Sollie P, Düwell M, eds. *Evaluating New Technologies.* Vol 3: Springer Netherlands; 2009:159-173. Available at http://dx.doi.org/10.1007/978-90-481-2229-5_11
60. Friends of the Earth U.S., International Center for Technology Assessment, ETC Group. *Principles for the Oversight of Synthetic Biology.*March 13, 2012. Available at http://www.foe.org/news/archives/2012-03-global-coalition-calls-oversight-synthetic-biology
61. EuropaBio. *GM crops: Reaping the benefits, but not in Europe. Socio-economic impacts of agricultural biotechnology.* 2011. Available at http://www.europabio.org/sites/default/files/position/europabio_socioeconomics_may_2011.pdf
62. Press Association. Europe must lift GM food limits to help feed planet, say experts. *The Telegraph.* October 27, 2014.

Available at http://www.telegraph.co.uk/news/earth/agriculture/11190795/Europe-must-lift-GM-food-limits-to-help-feed-planet-say-experts.html
63. David Cameron's science advisers call for expansion of GM crops. *The Guardian.* March 14, 2014. Available at http://www.theguardian.com/environment/2014/mar/14/scrap-dysfunctional-gm-regulations-uk-government-science-advisers-food
64. Plant scientists urge Europe to stop blocking GM trials on 'political' grounds. *The Guardian.* October 30, 2014. Available at http://www.theguardian.com/environment/2014/oct/30/plant-scientists-urge-europe-stop-blocking-gm-trials-political
65. Freedman DH. The Truth about Genetically Modified Food. *Scientific American.* 2013;309(3). Available at http://www.scientificamerican.com/article/the-truth-about-genetically-modified-food/
66. Nicolia A, Manzo A, Veronesi F, Rosellini D. An overview of the last 10 years of genetically engineered crop safety research. *Critical reviews in biotechnology.* Mar 2014;34(1):77-88. Available at http://www.ncbi.nlm.nih.gov/pubmed/24041244
67. Pomeroy R. Massive Review Reveals Consensus on GMO Safety. Real Clear Science. 2013. Available at http://www.realclearscience.com/blog/2013/10/massive-review-reveals-consensus-on-gmo-safety.html. Accessed September 13, 2016.
68. Specter M. Seeds of Doubt. *The New Yorker.* August 25, 2014. Available at http://www.newyorker.com/magazine/2014/08/25/seeds-of-doubt
69. The Editors of Scientific American. Labels for GMO Foods Are a Bad Idea. *SCIENTIFIC AMERICAN.* 2013;309(3). Available at http://www.scientificamerican.com/article/labels-for-gmo-foods-are-a-bad-idea/
70. Suresh A. *Kenya's maize famine underscores need for Africa to*

confront GMO fears. Genetic Literacy Project.December 8, 2014. Available at http://www.geneticliteracyproject.org/2014/12/08/kenyas-impending-maize-famine-underscores-need-for-africa-to-confront-gmo-fears/
71. Guguyu O. William Ruto says ban on GMOs to be lifted in two months. *Daily Nation.* August 12, 2015. Available at http://www.nation.co.ke/news/William-Ruto-ban-GMOs–lift/-/1056/2829368/-/wwxu2a/-/index.html
72. Mwaura G. Making case for genetically modified foods. *The New Times.* September 12, 2015. Available at http://www.newtimes.co.rw/section/article/2015-09-12/192467/
73. Cronin AM. *The Movement is Growing: Non–GMO Products Could Reach $264 Billion in U.S. Sales by 2017.* One Green Planet. 2013. Available at http://www.onegreenplanet.org/news/the-movement-is-growing-non-gmo-products-could-reach-264-billion-in-u-s-sales-by-2017/
74. Packaged Facts. *Non-GMO Foods: U.S. Market Perspective.*September 11, 2013. Available at http://www.packagedfacts.com/Non-GMO-Foods-7779884/
75. Whole Foods Market. *News release: Whole Foods Market commits to full GMO transparency.* 2013. Available at http://media.wholefoodsmarket.com/news/whole-foods-market-commits-to-full-gmo-transparency
76. Callaway E. Glowing plants spark debate. *Nature News.* June 4, 2013. Available at http://www.nature.com/news/glowing-plants-spark-debate-1.13131
77. Strom S. G.M.O. labeling bill clears first hurdle in Senate. *The New York Times.* July 6, 2016. Available at http://www.nytimes.com/2016/07/07/business/gmo-labeling-bill-passes-first-hurdle-in-senate.html
78. Strom S. G.M.O. labeling bill gains House approval. *The New York Times.* July 14, 2016. Available at

http://www.nytimes.com/2016/07/15/business/gmo-labeling-bill-gains-house-approval.html?_r=0
79. Thomas J. Ecover pioneers 'synthetic biology' in consumer products. *Ecologist.* June 16 2014. Available at http://bit.ly/Tgii0k
80. Barclay E. GMOs Are Old Hat. Synthetically Modified Food Is The New Frontier. *NPR.* October 3, 2014. Available at http://www.npr.org/blogs/thesalt/2014/10/03/353024980/gmos-are-old-hat-synthetically-modified-food-is-the-new-frontier
81. Colwell K. *News release: Haagen-Dazas says no to synbio.* Friends of the Earth.August 26, 2014. Available at http://www.foe.org/news/news-releases/2014-08-haagen-dazs-says-no-to-synbio
82. Lynch D, Vogel D. *The Regulation of GMOs in Europe and the United States: A Case-Study of Contemporary European Regulatory Politics.* Council on Foreign Relations.April, 2001. Available at http://www.cfr.org/agricultural-policy/regulation-gmos-europe-united-states-case-study-contemporary-european-regulatory-politics/p8688
83. Carter SR, Rodemeyer M, Garfinkel MS, Friedman R. *Synthetic Biology and the U.S. Biotechnology Regulatory System: Challenges and Options.* J. Craig Venter Institute. May 2014. Available at http://www.jcvi.org/cms/fileadmin/site/research/projects/synthetic-biology-and-the-us-regulatory-system/full-report.pdf
84. Regalado A. Why Kickstarter's glowing plant left backers in the dark. *MIT Technology Review.* 2016. Available at https://www.technologyreview.com/s/601884/why-kickstarters-glowing-plant-left-backers-in-the-dark/
85. Holdren JP, Shelanski H, Vetter D, Goldfuss C. *Improving Transparency and Ensuring Continued Safety in Biotechnology.*July 2, 2015. Available at https://www.whitehouse.gov/blog/2015/07/02/

improving-transparency-and-ensuring-continued-safety-biotechnology
86. Kuiken T. Shaping the future of synthetic biology. *Science.* Apr 17, 2015;348(6232):296. Available at http://www.ncbi.nlm.nih.gov/pubmed/25883347
87. Bagley MA, Rai AK. *The Nagoya Protocol and Synthetic Biology Research: A Look at the Potential Impacts.* Woodrow Wilson International Center for Scholars. 2015. Available at http://www.synbioproject.org/process/assets/files/6672/_draft/nagoya_final.pdf

AFTERWORD

PETER CARR

Sometimes what the world needs is a better balance between extremes—between trust and control, between belief and skepticism, between speed and safety. Dr. Gigi Gronvall's *Synthetic Biology: Safety, Security, and Promise* pursues a very important balance: between risk and opportunity. She first frames for us the tantalizing opportunities seen in the still-young field of synthetic biology. She also gives us a broad view of risks that come with those opportunities. We need to be well informed on both of these fronts in order to make the wisest choices.

There are certainly many applications of synthetic biology where the risks do not appear fundamentally different from the works of genetic engineering that have come before. In some cases there is a strong argument that newer tools for manipulating DNA—both more efficient and more precise—help reduce those risks. But synthetic biology also seeks out new territory that has only been lightly traveled before. There are opportunities to protect and enhance health with organisms engineered to live on and in the human body. There are opportunities to clean up chemicals in the environment and actively restore damaged ecosystems. There are even opportunities to engineer the cells of our own bodies to lower the chance of disease and enhance quality of life. In all these cases, the risks are easy to speculate on but hard to understand deeply.

For all of us, we need to be well informed to confront 2 untenable extremes in the ongoing debates over risks and opportunities from synthetic biology. On one side are strong voices for halting all progress, often citing fears that are comfortably "low resolution," broadly negative scenarios that are not substantiated with meaningful details. The precautionary principle is wielded as an absolute line drawn in the sand, but when a concern is addressed the line is simply redrawn elsewhere. To their opposition, this camp appears deeply invested in a no-risk fallacy, where no risk is acceptable, even though there is no such thing as no risk (including the risk from doing nothing).

At the other extreme, the voices can be rather quiet. Some innovators see the past decades of genetic engineering that are relatively incident-free, combined with their own experience in the lab (likely also incident-free), and generalize that the future outside the lab will be like the past (incident-free). "Leave us alone; we know what we're doing" is the impression they give to their opposition. That impression is reinforced by the many less-extreme scientists and engineers who privately acknowledge there may be meaningful levels of risk in some areas, but who hold back from public dialogue for fear of villagers with torches and pitchforks. They look at their colleagues who engaged in broader discussions of risk, who stuck their necks out only to have their heads bitten off, and say, no thank you. Furthermore, even the most brilliant synthetic biology researchers are still susceptible to "you don't know what you don't know" (sometimes *more* susceptible), including innovators unaware of how much they stand to learn from those who study risk professionally.

Both extremes have one seductive thing in common: They are easy. They allow one to stop thinking once reaching the conclusion of either "the chance of harm is clearly trivial" or "the worst outcomes I can imagine are clearly unacceptable." But most of us—technical elites and lay people alike—live somewhere between these 2 extremes. We understand that any truly powerful technology is subject to some level of accident or misuse and that we have a lot of thinking to do on the path forward.

This book seeks to help us find a better balance, and I encourage you to read it thoughtfully. The points Dr. Gronvall raises are crucial ones we need to consider as we chart a course that balances risk and opportunity. She highlights specific areas that merit deeper investigation, study, and discussion. I recommend that those involved in crafting government policy and funding research especially take note.

Please read and think. Then discuss and listen. And repeat. I will try to do the same.

Peter Carr, PhD, is a Senior Scientist at MIT Lincoln Laboratory.

(Opinions, interpretations, recommendations, and conclusions are those of the author and are not necessarily endorsed by MIT Lincoln Laboratory or the United States government.)

ACKNOWLEDGMENTS

This book would not have been possible without the support and encouragement of Dr. Paula Olsiewski and the Alfred P. Sloan Foundation. It is hard to overstate the value of Paula's leadership in addressing societal, ethical, and regulatory risks associated with synthetic biology, so that the field can advance in the public interest.

The written articles that shaped this work are referenced in the text, but the invaluable interviews and conversations conducted over the past several years with experts in synthetic biology, governance, national security, and biosafety, are not—so I would like to thank these experts for generously giving their time to share their thoughts and descriptions of their work, which influenced this book. In particular, I'd like to thank Spencer Adler, Gaymon Bennet, Kavita Berger, the late Roger Breeze, Roger Brent, Rob Carlson, Sarah Carter, Rocco Casagrande, Alison Cullen, Ian Dobson, Diane DiEuliis, Andy Ellington, Mike Elowitz, Maureen Ellis, Drew Endy, Jerry Epstein, Adam Finkel, Julie Fischer, Bob Friedman, Doug Friedman, Dave Franz, Michele Garfinkel, Laurie Garrett, John Glass, David Singh Grewal, Daniel Grushkin, David Guston, Jo Husbands, Mike Imperiale, Barbara Johnson, Ellen Jorgensen, Bob Kadlec, Gregory Kaebnick, Rebecca Katz, Jay Keasling, Alexander Kelle, Nancy J. Kelley, Jean-Claude Kihn, Tom Knight, Lori Knowles, Radha Krishnakumar, Todd Kuiken, Michael Kurilla, Jennifer Kuzma, Filippa Lentzos, Steve Maurer, Anne-Marie Mazza,

Gerald McKenny, Piers Millet, Allison Mistry, Susan Coller Monarez, Tom Monath, Maureen O'Leary, Tara O'Toole, Megan Palmer, Jason Paragas, Chris Park, Eleonore Pauwels, George Poste, Nicoletta Previsani, David Rejeski, Randy Rettberg, Alan Rudolph, Markus Schmidt, Fran Sharples, Pamela Silver, Bernard Silverman, Amy Smithson, Erin Sorrell, the late Pim Stemmer, Joyce Tait, Terry Taylor, Tim Trevan, Andy Weber, Jim Welch, Carrie Wolinetz, Jaime Yassif, and Laurie Zoloth. Many of these interviews took place while I was in the process of working on research projects funded by the Alfred P. Sloan Foundation, which supported the research and writing of this book, as well as projects funded by the Defense Threat Reduction Agency and the Council on Foreign Relations.

I have some great colleagues who read many sections of the manuscript and offered terrific advice: Tom Inglesby, Matt Watson, and Crystal Boddie. Thanks also to Michelle Rozo, Tara Kirk Sell, Amesh Adalja, Anita Cicero, Andrea Lapp, Hannah Collins, Matt Shearer, and Tanna Liggins for help along the way, including research that contributed to the analyses herein, and to Sanjana Ravi, Ryan Morhard, Kunal Rambhia, and Mary Beth Hansen for work we did together on related projects funded by the Sloan Foundation. Jackie Fox provided invaluable suggestions, skilled and last-minute editing, and help throughout the manuscript development and publication. Davia Lilly created the cover design. Thanks also to Yolanda Wolf at Sloan for keeping this project running. I, of course, own all the errors in this book.

Randy Larsen and Ken Bernard not only read major portions of the manuscript and offered advice, they have been wonderful, inspiring mentors to me—as always, I am so grateful for their help. And of course, I am indebted to DA Henderson, who I—and so many others—will miss a great deal.

I'd like to give a big thank you to my sister, Karen, and to all my friends in South Baltimore, especially the friends who are also family: the Stewarts and Kopecky/Henry's. And every day I thank my lucky stars for the support and love of my husband, Jesper, and our two boys, Casper and Felix, who we love more than anything.

INDEX

www.ingramcontent.com/pod-product-compliance
Lightning Source LLC
Chambersburg PA
CBHW070235190526
45169CB00001B/193
9781539336839